A BRIEF HISTORY OF THE HUMAN RACE

ALSO BY MICHAEL COOK

Forbidding Wrong in Islam: An Introduction

The Koran: A Very Short Introduction

Muhammad

A BRIEF HISTORY OF THE

HUMAN RACE

Michael Cook

Granta Books
London

Granta Publications, 2/3 Hanover Yard,
Noel Road, London N1 8BE

First published in Great Britain by Granta Books 2004
First published in the US by W. W. Norton & Company, Inc. 2003

A CIP catalogue record for this book
is available from the British Library.

3 5 7 9 10 8 6 4 2

ISBN 1 86207 687 1

Printed and bound in Great Britain
by William Clowes Limited, Beccles and London

For Margot

who was born in China, and migrated to Colorado,
while this book was being written

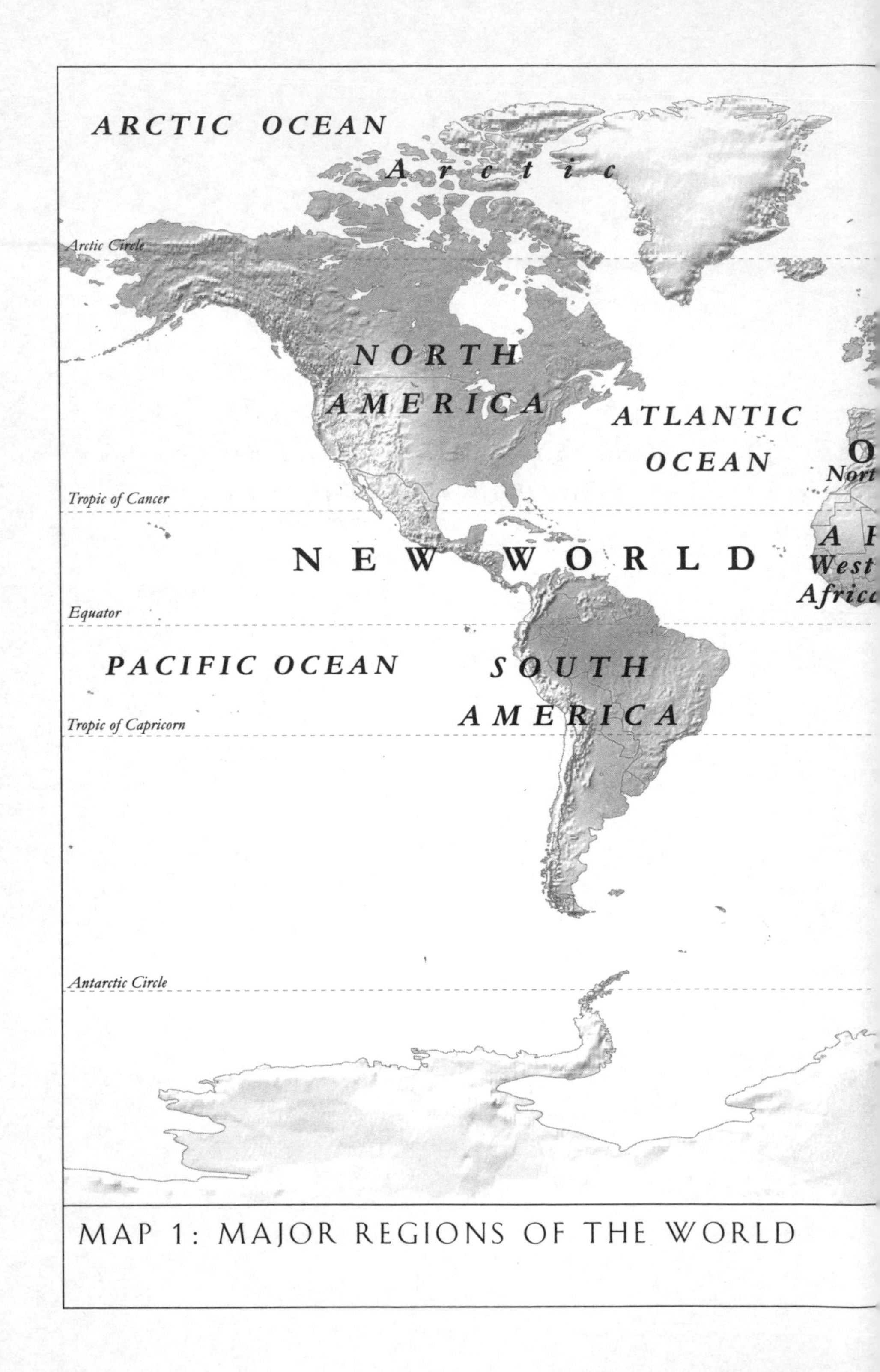

MAP 1: MAJOR REGIONS OF THE WORLD

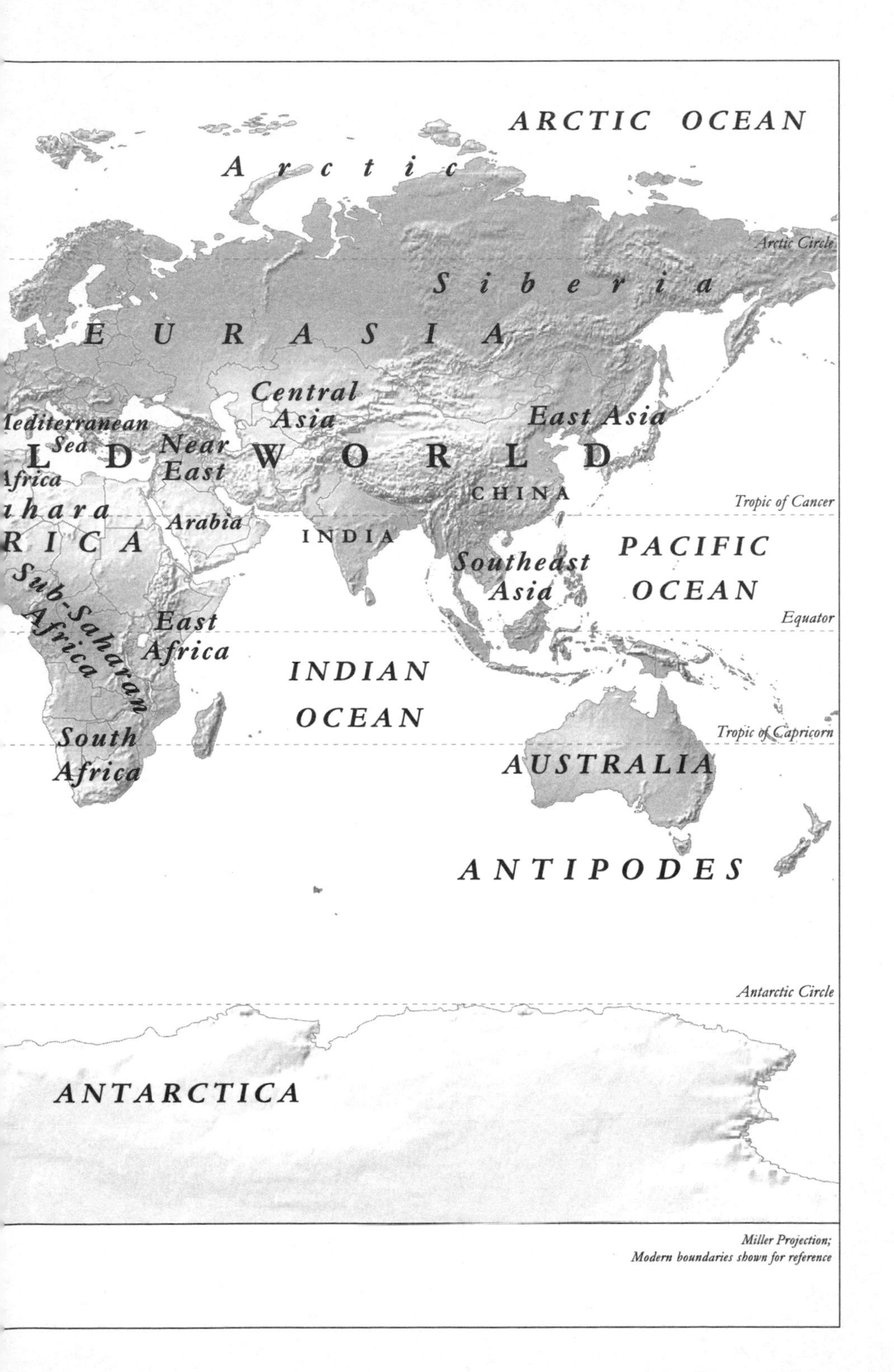

Miller Projection;
Modern boundaries shown for reference

CONTENTS

PART TWO
THE SMALLER CONTINENTS

PART THREE
THE EURASIAN LANDMASS

LIST OF MAPS

LIST OF FIGURES

PREFACE

Humans have taken to making history only in the last few hundred generations. For two or three thousand generations before that our ancestors were probably no less intelligent and insightful than we are now (or no more stupid and obtuse). But they were otherwise engaged. Why they changed tack, and with what results, is what this book is about.

There are at least two very good reasons not to write a book like this about history. One reason is that there is so much of it. A few hundred generations may not be a very long time in biological terms, but when we are dealing with the human race they add up to a surprising amount of cultural change; what is more, our activities have been spread over several continents. The result is that the information we possess about our past fills libraries. No one can know all there is to be known about it, let alone hope to convey even the gist of it in one small volume.

The other reason is that there is so little of it. Most of the history made by our ancestors is irretrievably lost to us. Despite the abandon with which humans litter the planet with bones, pot-

Fig. 1: Three objects from ancient Greece.

sherds, atomic waste, memoirs, and the like, the fact remains that most of our history is dark matter, and what is not often seems too fragmentary and obscure to repay much attention. Consider, for example, the three clay objects in figure 1. They form a small part of the litter left to us by the ancient Greeks, who must have made them for some purpose; but what could it have been?

The result is that this book is both deliberately selective and involuntarily patchy. As Voltaire said somewhere, "the secret of being boring is to say everything," and that at least is a trap I have avoided falling into. There is, of course, a fine line between making things too simple to be informative and making them too complicated to be assimilable. I hope I have found my way along it, but that is something for the reader to decide. If my book does not work for you, you can turn to any number of others with different approaches.

I should emphasize that in the end this book is intended to

serve a distinctly humble purpose. It seeks only to convey to an alert reader an overall sense of the shape of human history and an idea of some of the ways in which it is interesting. By this I mean that I do not have a Grand Unified Theory of history to offer. This, in fact, may be no great loss, since if I had such a theory, it would almost certainly be wrong. I do have ideas. Some of them are the published ideas of others, and some are my own hitherto unpublished ones. But they are not ideas that stand or fall together; while it is likely enough that some of them are wrong, there is also a good chance that some are right.

The structure of the book as a whole will be evident from a glance at the table of contents. In particular, it should be obvious that this is not a book for which world history is modern history with a short prologue. Admittedly there is an element in my thinking of "how we got to where we are now," if only because I have no idea where we will get to in the future. But I have tried to resist the hollow pretensions of the present to be the chronological center of the universe.

What may not be quite clear from the table of contents is the makeup of a typical chapter. The long section with which each chapter begins provides a broad survey, interrupted by a few subheadings here and there. Then follow two shorter sections, each of which has a much narrower focus. These are close-ups. Often the second section of the chapter takes up a particular theme, while the focus of the third is on some class of objects, in many cases of a kind that can easily be seen in museums; but this is not always so. In selecting these close-ups, I have tended to pick topics that I personally find interesting or significant; no two authors following this recipe would choose the same ones. Some of these sections are about how we know things we think we know about the past. Others pick up characteristic and sometimes extravagant features of the cultures under discussion, or look at the things cultures can get up to if they have access to their distant pasts. I have also used these sections to redress the balance in favor of some topics that get shortchanged in the survey sections. Women play a role here, as does science (both as an activity of past societies and as a source of information about them).

I first found myself teaching an approximation to world history at the School of Oriental and African Studies in London in the late 1970s and early 1980s. The course was on premodern Asia and Africa; my colleague Colin Heywood took Asia from the third century A.D., and I was left with the rest. My interest in many of the issues discussed in this book dates from that distant epoch, as do several of my ideas about them. I then moved to Princeton, and did not return to the subject until the late 1990s, when my interest in it was rekindled by reading Jared Diamond's *Guns, Germs, and Steel.* I then taught a course on world history to small groups of Princeton undergraduates, and thanks to them I became thoroughly reimmersed in it; their insights and arguments have affected the way I think about the subject.

But this hardly explains why I should write a book. I would not like to be thought of as a professor who cannot teach a course without broadcasting it to the world at large. Perhaps a couple of things were crucial in pushing me over this rather tempting edge. First, it was in 2000 that I finally published a seven-hundred-page, elaborately footnoted monograph on a topic in my specialist field that I had been working on for fifteen years. This freed up the time for me to write about something else, and at the same time delivered me from any immediate sense of obligation toward my field. Second, I had long been talking to Steve Forman, whom I knew thanks to Robert Wisnovsky, about a book on Islamic history that I may yet write some day for Norton. When I mentioned my course on world history, he suggested a book on it. This is the book.

A word on conventions.

When I need to make clear whether a date falls before or after the turn of our era, I use the traditional forms "B.C." and "A.D.," rather than the more recent euphemisms "B.C.E." and "C.E." I do not thereby affirm the messianic status or divinity of Jesus of Nazareth, just as I speak of "January" and "Wednesday" without committing myself to the worship of the Roman god Janus or the Germanic god Woden. The fact that the era most widely used in the world today is historically a Christian one says something about the course of world history, and it is not a course that everyone has reason to celebrate. But righting the wrongs of history by euphemism seems at best a somewhat idle pursuit.

For transcribing Chinese names and terms I use the Wade-Giles system rather than Pinyin, but I supply Pinyin equivalents in parentheses at the first occurrence. An eminent historian of China recently described both systems as abominable; I have chosen Wade-Giles because the mispronunciations it engenders for the average reader of English are significantly less bad than those encouraged by Pinyin.

I use the word "farming" as a general term covering the cultivation of domesticated plants (agriculture in the narrow sense) and the tending of domesticated animals (mainly pastoralism). I am not entirely happy with it, since it is a little too closely linked with the history of English land tenure; but the more precise, if inelegant, term "food production" has not really caught on.

For critical comments and suggestions, I would like to thank Patricia Crone, who read the typescript for me, and William Everdell and John E. Wills Jr., who did so for Norton. I also profited from the comments of my sons Simon and Richard. David Shulman helped me with south Indian matters; Svat Soucek assisted me with an archaic Russian text. Over the years many people must have told me or given me things that have found their way into this book; they include at the very least John Brinkman, Michael Doran, Oleg Grabar, Kenneth Mills, Stephennie Mulder, and Frank Stewart. Şükrü Hanioğlu, Giovanna Cesarani, Zilan Shen, and Leon Zhu gave me practical help with some of the illustrations. Sarah England handled scores of problems arising from the illustrations and maps with energy, precision, and good humor. I have also derived much benefit from the kindness of the staff at the Princeton University Art Museum, and Michael Padgett in particular, in hosting classes during which my students and I were able to handle artifacts of the cultures we were studying. I am glad to say that we did not simulate the ravages of time by actually breaking any of them.

PART ONE

WHY IS HISTORY THE WAY IT IS?

CHAPTER 1

THE PALAEOLITHIC BACKGROUND

1. Why Did History Happen When It Did?

Figure 2 is taken from a photograph that appeared in the *New York Times* in 2001. Unless you know more than my caption tells you, it is not particularly exciting; but it is worth looking at as a record of human activity. You should have no difficulty in identifying the general character of the scene. Some people, presumably archaeologists, must have recently excavated the ruins of old stone buildings erected by other people, presumably non-archaeologists who once lived in them. But going beyond this description of generic ruins excavated by generic archaeologists is a problem. For example, if you ask yourself where these ruins are located, your chances of coming up with even the right continent are not much better than one in six; indeed, the only continent you can unhesitatingly exclude is Antarctica. This geographical uncertainty is an interesting point in itself, and we will come back to it in a later chapter. But for now, let us turn instead to the question *when* the walls could have been built. Here you are hardly better off. You can take it that

Fig. 2: Recently excavated ruins.

the ruins must date back at least a century or two, or they probably would not be so ruined, and the archaeologists would not have bothered to excavate them; also there is no sign of modern building methods. But how far back could these ruins date? If you have a general notion of the chronology of the human past, you will be reasonably confident that they cannot be older than about ten thousand years, and that within that period they are more likely to date from the last few thousand years than from the first. As a dating, that seems terribly vague.

Vagueness, however, is in the eye of the beholder. Modern humans—in the sense of people anatomically indistinguishable from us—date back a good 130,000 years, and perhaps considerably longer. Against that background, a dating of the ruins to the last 10,000 years seems strikingly precise; 10,000 years is, after all, no more than a few hundred generations. Why, then, should humans have been around for so long before they thought to erect something as simple as a building with stone walls? Why would it be a major sensation if these ruins turned out to be, say, 30,000 years old?

The practice of building stone walls is not in itself a central theme of this book. Instead, I have been using it as a proxy for a much wider range of human activities that we can sum up as making history. There is a conventional definition of history as that segment of the human past that is accessible through written sources; on this criterion, history dates back no more than five thousand years. This is not a frivolous definition. Given the centrality of language in human cultures, it makes a great difference to us whether or not we can study past humans on the basis of what they had to *say* for themselves; and given the potential uses of writing in human society, it made a considerable difference to the peoples themselves whether or not they possessed this fundamental item of information technology. But to start history five thousand years ago is to break into the middle of a story of rapid cultural change that began another five thousand years earlier, with the emergence of farming. In this sense the last ten thousand years form a well-defined segment of the human past that stands in marked contrast to the much longer period preceding it, a period in which hunting and gathering were the only way of life practiced by humans. So for convenience, let us decide to call these last ten thousand years of the human past "history." We can now replace our question about the ruined walls with the one that stands at the head of this section: Why did history happen when it did? Why has it all been packed into the last ten thousand years?

A simple question deserves a simple answer, and this may be one of the rare occasions on which it actually gets one. Figure 3 shows some results from an ice core drilled in Greenland in the early 1990s. The oldest ice takes us back a quarter of a million years (at the far right), the newest brings us virtually to the present (at the far left). Because ice thins with age, we get better resolution for more recent times; this is why, in chronological terms, the scale

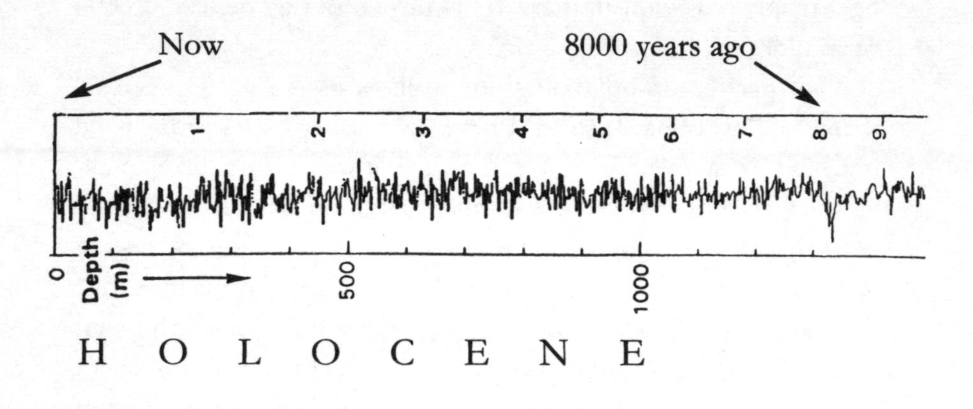

Fig. 3: Data from the Greenland ice core.

is larger toward the left. What the data show is the movement of a ratio relating two distinct isotopes of oxygen. This ratio interests us because it serves as a stand-in for temperature: up is warmer, down is cooler. Data from other sources give us a fair confidence that these temperature movements say something about changes in global climate (and not just local conditions in Greenland). So what do we see here, apart from unending wiggles?

There are really only two points we need to note, but taken together they bid to give us a large part of the answer to our question. The first point is that the last 10,000 years—the period geologists call the Holocene—has been quite unusually warm. To find a comparable time in the late Pleistocene, the geological epoch preceding the Holocene, we would have to go back to the Eemian period some 120,000 years ago. This in fact is fairly typical of the pattern of the last million years: relatively short warm periods recur every hundred thousand years or so. The second point is that the Holocene has been privileged in a further respect: its extraordinary climatic stability. Compare the gentle wiggles on the left with the wild lurches on the right. (Indeed, if we look again at the Eemian, it appears to be made up of a series of peaks and valleys that were respectively far hotter and far colder than anything we

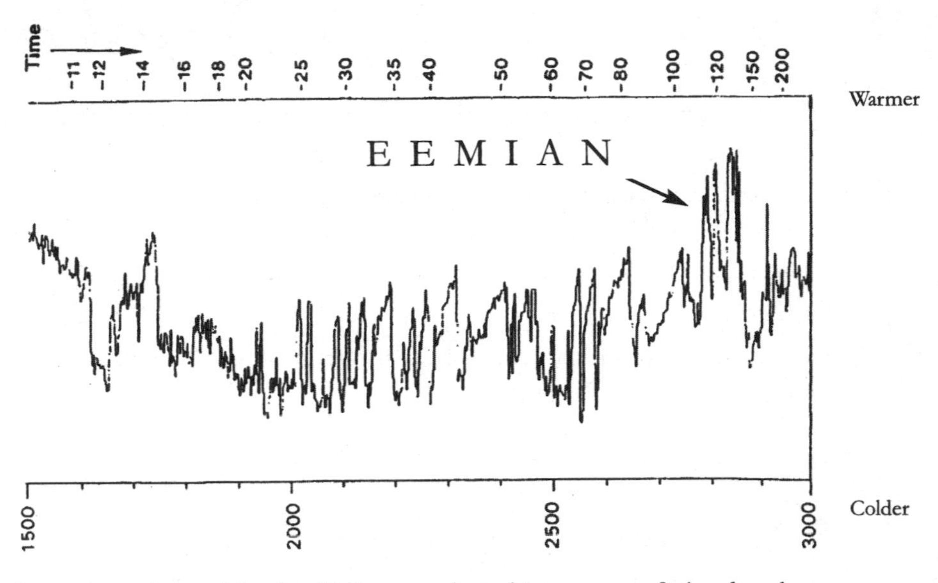

have experienced in the Holocene; but this aspect of the data has been called in question.) The Holocene is thus a very unusual period. There has been nothing like it in the last 100,000 years, which is most of the period during which modern humans are likely to have existed.

History, as we have defined it above, fits snugly into the warm and stable climatic niche of the Holocene. This association hardly seems mysterious. Trying to make history in an ice age—which is most of what there was between the Eemian and the Holocene, and before the Eemian—would not have been much fun. More specifically, human history is founded on farming, and it makes good sense to suppose that its development and maintenance would have been very difficult, if not actually impossible, in a world climate that was not both warm and stable; certainly no trace of farming has yet been found in the Pleistocene. The Holocene, then, was the window of opportunity for the making of history. If this is right, there is no reason to ask why humans should have waited so long before making history; if we leave aside the Eemian, it seems that they jumped through the window just about as soon as it opened for them.

That gives us one big part of the answer to our question. But we still need something else, for the Holocene window could have opened and closed without anyone being there to jump through it. After all, those species most closely related to humans—chimpanzees, gorillas, gibbons, orangutans—did not respond to the Holocene by developing farming; we did, and with extraordinary consequences. A serious answer to the question why we were different would take us too far afield. But there is something about the general character of modern humans that can help explain their response to the Holocene.

The statement made above to the effect that people anatomically indistinguishable from us date back 130,000 years or more is a statement about fossils. Fossils typically reveal gross anatomical features, above all those relating to bones. A different kind of evidence about early humans is provided by archaeology, which unearths the artifacts that they leave behind them. We will take a look at some of these artifacts later in this chapter. What matters here is a point about their chronology. If we go by the archaeological record, we find no reason to place the emergence of modern humans—in the sense of behaviorally modern humans, with innate capacities we can take to be similar to ours—as far back as 130,000 years ago; 50,000 would be more like it. How these different datings can be reconciled is a serious problem in the study of human evolution. Either the anatomically modern humans of earlier times possessed the same capacities as we do, but did not use them in ways that left extensive archaeological traces until 50,000 years ago; or the behaviorally modern humans of the last 50,000 years were the result of a biological change—perhaps linked to some stage in the evolution of language—that left no trace in the fossil record. (The prehistory of language has been one of the most frustrating blanks in our knowledge of the origins of our species.) Fortunately, for our purposes it is enough to identify the problem; we do not have to solve it.

Instead, let us focus on what is distinctive about the archaeological record of these behaviorally modern humans. It is not just that their culture appears to have been significantly richer than anything previously attested—most famously, we now have clear examples of sophisticated art, thanks to rock paintings well over twenty thousand years old found as far apart as Europe, South Africa, and Australia. We also find evidence of a new cultural diversity: artifacts show unprecedented cultural variation over both space and time, a point we will return to later in this chapter. So what we see, against the background of the earlier record, is not just more culture but a well-developed capacity for cultural change. It is not hard to see why this endowment could have been adaptive in a period of violent climatic swings; cultural agility is one way to respond to rapid environmental changes. This same agility was of fundamental importance in the human response to the Holocene.

So here we have a plausible answer to one basic question about history, the question why it happened when it did. On the one hand, the Holocene was an exceptional, perhaps unique window of opportunity in the lifetime of behaviorally modern humans. And on the other hand, these humans possessed a unique cultural agility with which to respond to the opportunity.

II. Genetics and the Origins of the Human Race

The formidable Muslim scholar Ṭabarī (d. A.D. 923) has left us a massive history of the world as he knew it. Typically all he does is quote earlier sources, but near the beginning of the work he makes a significant methodological statement in his own voice: "no knowledge of the history of men of the past and of recent men and events is attainable by those who were not able to observe them and did not live in their time, except through information and transmission provided by informants and transmitters." For anyone who does not get it, he adds that "this knowledge cannot be elicited by reason or inferred by internal thought processes." Over most of human history, most of the time, Ṭabarī is right. But in modern times we have started to make serious use of various kinds of nontextual evidence, sometimes to supplement the texts and sometimes to compensate for their absence. Thus in the preceding section, for which we have no textual evidence whatever, the discussion was dominated by two bodies of evidence that Ṭabarī either did not know or ignored: fossils and artifacts. In recent years these have been joined by a third, which Ṭabarī could scarcely have dreamt of: genetics. What genetics has to tell us is sufficiently remarkable to be worth a closer look. But before we go on to the results of these genetic studies, we should run over some of the basic concepts involved.

The key element in the human genome is a set of instructions for building and running the human body. These instructions are made up of some tens of thousands of genes. The genes are written in what is known as the genetic code, a four-letter biochemical alphabet of which three letters make a word; the medium in which the message is written is a well-known biochemical substance, DNA. This material, in turn, is assembled in the form of twenty-three chromosomes, of which two sets are typically found in the nucleus of each cell of the human body. A complete transcription of the genome of a human being, as found in any one cell, would fill a small library. But curiously enough, about 99 percent of this text has been thought to be of no functional significance; such nonfunctional material is known as junk DNA. The remaining text, the part that matters, might fill an encyclopedia. We need not bother ourselves with the way in which these instructions are carried out; what matters for our purposes is that the genome is hereditary. More precisely, your genome is made up in a rather random way of some segments taken from your mother's genome and others taken from your father's. This at least is how it is in general; but two exceptions to this randomness will concern us in this or later chapters. One is the Y chromosome, which makes a male; only a male can have one, and he necessarily inherits it from his father, and so on back to Adam. The other is something known as mitochondrial DNA. Mitochondria are organelles found in considerable numbers outside the nucleus of the cell; they would seem to have originated as distinct organisms, but they have been living symbiotically inside cells ancestral to ours for the last billion years or so. One vestige of their independent origin is that they retain some DNA of their own. What matters here is that you inherited your mitochondrial DNA from your mother, who had it from her mother, and so on back to Eve.

One thing that has to happen when DNA is inherited is that it is copied. Copying texts is something of which we all have experience: it leads to mistakes, and the more successive copies are made, the more mistakes accumulate. Some of these mistakes may be disastrous, some may be no more than a nuisance, and some will not matter at all. In the case of DNA, mistakes in copying are called mutations. The overwhelming majority of mutations do not matter, because they fall in the junk DNA (whereas mutations in genes can cause unpleasant diseases like cystic fibrosis). But mutations that do not matter may suit our historical purposes very well, because they are simply passed on; there is no natural selection operating for or against them.

By and large, DNA deteriorates rapidly once it is no longer part of a living cell. This means that in general we can study only the DNA of the living, or of the recently dead when their DNA has been preserved in favorable conditions. The DNA of humans alive today is thus fairly readily available, as is that of other living species more or less closely related to us. But usable DNA more than a few thousand years old is a rarity. Curiously, we currently possess no fewer than three samples of the mitochondrial DNA of Neanderthals, the premodern humans who lived in Europe from about 130,000 to about 30,000 years ago. But we have no assured samples of the same vintage for modern humans, and nothing for earlier hominids.

How, then, do such mutations enable us to reconstruct the past? Rather than try to grasp this in the abstract, let us go straight to a question they shed light on: the origins of the human race. In the last few decades, there has been a noisy dispute about this. According to one camp, the origins of modern humans are multiregional: that is to say, the process by which we emerged was one in which a variety of premodern hominids in different parts of the world (among them the Neanderthals) evolved in the same direction. According to the other camp, modern humans evolved in a single place—somewhere in Africa—and spread from there to the rest of the world, replacing earlier hominids (such as the Neanderthals) in the process. The controversy got under way on the basis of the fossil evidence, but it was not resolved by it. Significant genetic studies began in the 1980s, and by now it seems clear that they have shown the "out of Africa" school to be right. We can sum up the results of these studies in three major points.

First, it turns out that, from a genetic point of view, the human race is remarkably homogeneous. That is to say, we are much more genetically alike than are the various species of ape to which we are most closely related. This is a politically welcome conclusion in the world today: it means that if you are a racist, in the sense that you would like to see deep genetic cleavages within your own species, you would really be much better off as a chimpanzee (a species within which zoologists have no trouble identifying three distinct subspecies). But the benign character of this finding should not distract us from its interest. Things did not have to be this way, and the fact that they are this way must indicate something about our past: most obviously, that the human race is too recent in origin for enough mutations to have accumulated to make us as differentiated as the chimpanzees.

Second, genetic studies do something to quantify the word "recent." The basic idea here is to calculate how long it must have taken for the observed number of mutations to accumulate. This, of course, involves questions of calibration and statistical method that we must leave aside. To date the answers vary widely, but are mostly within the range of 50,000 to 200,000 years ago. This is comforting to the extent that it indicates that the figures we were talking about on the basis of the fossils and artifacts are of the right order. But it is frustrating in that as yet it does not help us to decide whether the human race as it exists today has been diverging genetically for as much as 130,000 years (as would doubtless be the case if the humans of that time were already fully modern) or as little as 50,000 years (as would surely be the case if fully modern humans appeared only then).

Third, the genetic evidence points more and more confidently to Africa as the birthplace of the human race. What is involved here is a very general principle applicable to anything (a species, a language, a cultural practice) that originates in one region, spreads to others, and is subject to mutation over time. Other things being equal, such a phenomenon should be most differentiated in its ancient homeland, for the simple reason that it has been there longest; by contrast, it should be least differentiated in the region to which it has most recently spread, because it is likely to have arrived there in one particular form that will not yet have had much time to break up into distinct varieties. Now studies of mitochondrial DNA show that African populations are indeed the most deeply differentiated both from each other and, in general, from non-African populations; non-African populations bunch together with a single group of African populations. The simplest and most obvious interpretation of these results is that the human race originated and began to differentiate in Africa, and spread to the rest of the world only at a significantly later stage.

This leaves us with an obvious question that takes us away from genetics: Why Africa? The question is larger than it looks. We have good reason to think of Africa as the site of the emergence not just of modern humans but of the entire hominid line that eventually led to them; indeed, there is no central event of hominid evolution over the last few million years that we can confidently place outside Africa. What, then, was so special about Africa? It is easy enough to eliminate the Americas and Australia from the running: they lacked apes, and the first hominids to reach them were modern humans. But why Africa rather than Europe or Asia, given that apes existed in all three?

A full answer to this question would take us well beyond the confines of this book, but one rather general point is worth making. Africa is strong on tropics. This matters because it is in the tropics that the input of energy from sunlight is greatest. The highest density of species anywhere on earth is thus found around the equator; as we move north or south from the tropics, the numbers fall off dramatically. The result is that, other things being equal, the appearance of new species is significantly more likely in the tropics than elsewhere. This in turn suggests a question for us to take up later: Why did the high-profile role of the tropics in the making of the human race fail to carry over into the making of history? In the meantime let us shift back to some evidence of the human past that, unlike DNA, is at least visible to the naked eye.

III. Stone Tools

If an interest in early humans takes you to a museum, most of what you see is stone tools. Indeed, hominids have been making and using such tools for two and a half million years; only in the last few thousand have they moved on to metal. The period between the first appearance of stone tools and the first appearance of metal tools is known as the Stone Age (the term was coined by a Danish museum director in the 1830s, along with the terms Bronze Age and Iron Age). But this usage is unwieldy, and begs for subdivision. So we first separate the Old Stone Age, or Palaeolithic, from the New Stone Age, or Neolithic (*lithos* being the Greek for "stone"); we can set the boundary at the emergence of farming, which in the Near East means around ten thousand years ago. Then we further subdivide the Palaeolithic into Lower, Middle, and Upper. It is the Upper Palaeolithic that is the focus of our interest in this chapter, covering the period from the definite appearance of modern human behavior some fifty thousand years ago to the emergence of farming.

This terminology of stone ages is convenient, but it needs to be put in its place. An obvious objection to it is that we have not entirely abandoned stone tools in the last few thousand years, or even in modern times; in the Near East, for example, they were still in use in the twentieth century for a variety of purposes, from trimming toe nails to cutting umbilical cords. A slightly less obvious problem is that stone tools are unlikely to be representative of the tool kits of Stone Age societies; they are simply what survives best. Most tools most of the time are likely to have been made of wood, but the conditions that preserved a 300,000-year-old wooden spear at Clacton in England are quite exceptional. Nevertheless, we can be sure that stone tools were at least a significant part of the tool kit of the societies that used them: making stone tools takes a lot of work, and our ancestors would hardly have kept at it had it not offered them considerable rewards.

To make a stone tool, the first thing you need to do is pick the right kind of stone: hard, fine-grained, and isotropic (not predisposed to split in some planes rather than others). Generally the kind of rock you want will be igneous; flints tend to be good, but they are by no means the only possibility. Next, you need to break off workable flakes. Figure 4 shows one way to do this as observed in an effectively Neolithic culture in New Guinea. The man is using the stone he holds in his hands as a hammer to strike flakes off the boulder between his legs. His posture has the advantage that flakes and other debris from the boulder will tend to fly out behind him; making stone tools is risky work, particularly if a chip hits you in the eye. Once you have your flakes, the next stage is to shape them. Figure 5 shows men of the same New Guinea village at work on this: the basic technique is to hold the flake in your left hand, and strike it with another stone held in your right hand. As might be expected, it takes a lot of experience to get it right.

The uses of stone tools are by no means apparent to people

Fig. 4: A New Guinea stone worker quarrying flakes in 1990.

who no longer use them. It is telling that archaeologists have had considerable difficulty distinguishing the earliest such tools from the waste products associated with their production; and many things that are called hand axes probably weren't. But observation of recent hunter-gatherers has shown the purposes for which they typically use stone tools: to crack nuts, chop or shape wood, cut up meat, work hides, and fight each other. This at least gives us an idea of what to look for. Putting stone tools under a microscope and examining them for patterns of "microwear" has confirmed that they were used for butchery and woodwork as early as one and a half million years ago. A close look at the Clacton spear shows that it was likewise shaped with stone. And as to fighting, we do not need forensic medicine to interpret the occasional sharp stone found lodged in a Palaeolithic skeleton.

When we raise our eyes—still hopefully intact—from these stones, what have we learned? Let us begin with the phenomenon as a whole.

The first point to note is that we have here a pattern of tool use that is confined to relatively advanced hominids (to members of

Fig. 5: New Guinea stone workers shaping their flakes.

the genus *Homo*, as the experts put it). This sets us apart from our earliest hominid ancestors, and also from our cousins among the apes. Some chimpanzee populations in the wild use stones to crack nuts (just as most of us have done on occasion when we did not have a metal nutcracker to hand); yet they do not *work* stones for this or any other purpose. Experiments have shown that a captive pygmy chimpanzee can learn to make stone tools to cut string, though at a level well below that of the oldest hominid work; but this behavior was induced by the human experimenters, and is not found in the wild. In the same way the making of stone tools sets later hominids apart from earlier ones—there is no good evidence that the small-brained Australopithecines had such a technology.

The second point is that this uniquely hominid practice is culturally transmitted (just like chimpanzee nutcracking). Or at least, we know for a fact that modern humans are not genetically programmed to make stone tools. They usually learn how to do it from other humans, if at all, and earlier hominid makers of such tools are unlikely to have been different in this respect.

The third point is that, despite the presumably cultural transmission of the technique, the lithic record tends to be remarkably stable until the Upper Palaeolithic. There are changes and differences, but most of the time techniques seem to be pretty much invariant over distances of thousands of miles and time spans reaching hundreds of thousands of years. Hence the coarse-grained character of the labels used by European archaeologists for the cultures of those times: "Acheulian" covers many hundreds of thousands of years, "Mousterian" many tens of thousands.

By contrast, the Upper Palaeolithic, as we have seen, marks the onset of cultural diversity. In quick succession, the same archaeologists identify the Aurignacian, Gravettian, Solutrean, and Magdalenian cultures, with durations of no more than 4,000 to 12,000 years apiece. Thus stone tools of the Aurignacian (about 40,000 to 28,000 years ago) and Solutrean (about 21,000 to 16,500 years ago) look distinctly different. In the same way cultures now varied from one region to another. So the stone tools reveal to us the emergence in the Upper Palaeolithic of a new cultural mutability. What happened when this mutability was conjoined with the climatic opportunity of the Holocene is the subject of the rest of this book.

CHAPTER 2

THE NEOLITHIC REVOLUTION

I. Why Has the Making of History Been a Runaway Process?

In the late twentieth century humans developed a new respect for chimpanzee culture. A survey published in 1999 listed no fewer than thirty-nine items of chimpanzee behavior in the wild that seemed to be culturally transmitted. For example, as we saw in the preceding chapter, some chimpanzee populations have a technique for cracking nuts with stones—more precisely, they place the nut on one stone and break it with another. Chimpanzees are not genetically programmed to do this; all they get from their genes is the capacity to acquire the skill, normally by learning it from other chimpanzees. This kind of evidence makes it abundantly clear that humans are not the only animals that have culture.

Yet by human standards, the culture of chimpanzees is very limited. In the first place, they do not seem to accumulate it to any significant degree. Of course, our knowledge of their culture is confined to what we now observe (chimpanzee nutcracking was first described in 1844); we can only guess what it was like two thousand, or two million, years ago. But if what we see now is all they have accumulated over an indefinitely long period, then by our standards it is not very much. In the second place, their culture does not seem to be of the kind in which one thing leads to another. Humans domesticate the horse and invent the wheel; they put them together and have a horse and cart; eventually they replace the horse with an internal combustion engine and have a car. In the same vein we might imagine nutcracking chimpanzees substituting a third stone for the nut and shaping a stone tool; this is something hominid toolmakers did, in a slightly more complicated version of the technique described in the preceding chapter. But chimpanzees do not do this, and in general it does not seem that one item of their culture serves as a platform for the development of another. In sum, just as chimpanzee culture is not cumulative, so also it is not dynamic.

Human culture, at least potentially, is both cumulative and dynamic. But this potential is unlikely to be taken very far if humans are constrained to lead a simple life, as hunter-gatherers usually are. This is most obvious at the level of material culture. Hunter-gatherers are typically nomadic because their mode of exploitation of their environment is extensive rather than intensive. Like sensible travelers today, nomads travel light: if you acquire a new artifact, you had better dump an old one. At the same time, having a restricted cultural inventory greatly reduces the chances of one thing leading to another. These arguments should not be pressed too far. Some lucky hunter-gatherer populations have inhabited niches so rich that they have been able to dispense with nomadism. One possible example is the European reindeer hunters of the Magdalenian culture, in the last millennia of the Upper Palaeolithic. It is perhaps connected to this that the last stages of pre-Neolithic culture in some regions seem to be appreciably richer than what went before them, to such an extent that the archaeologists have created for them the intermediate category of the Mesolithic. But by and large, we can see Palaeolithic hunter-gatherers as precluded from realizing the full potential of human culture by the constraints of the relationship between their way of life and the environment they lived in—which is not to say that realizing this potential is necessarily a good idea.

The Emergence of Farming

It is with the emergence of farming that these constraints are definitively lifted, and the runaway process of making history sets in. Why this should be so is not perhaps immediately obvious. We could easily imagine a world in which farming made an appearance, sputtered, and died out, or one in which it stuck, but led to no further innovations of consequence; if there were archaeologists in such worlds, they would hardly be tempted to honor the emergence of farming by dubbing it the Neolithic revolution. But in our world it really did amount to a revolution, and we need to understand why.

The story begins in the Near East in the ninth or maybe the tenth millennium B.C. Why then and there? The archaeologists have quite a lot to say about the immediate background to the process, and exactly what may have triggered it. But for our purposes, a broader perspective may be more helpful. At this level the question "Why then?" need not detain us, because we already know the answer: we are looking at the beginning of the Holocene. The question "Why there?" is more interesting. It comes down to a point so obvious that it is rarely put into words. The basis of farming, and hence of the whole historical development of human societies, is grass. This may sound counterintuitive, since for most of us grass is the harmless stuff of which urban parks and suburban lawns are made. But to see the truth of it, consider the fact that farming is a package with two major components: the cultivation of domesticated plants, and the tending of domesticated animals. Among the plants, the oldest and still by far the most important are the domesticated forms of grass we know as grain crops—such as wheat and barley. Among the animals, the analogous position is occupied by herbivores—the sheep and cattle on which pastoralists depend for a living. This immediately explains why the tropics have taken a back seat in the course of history (even today, underdevelopment has an elective affinity for the tropics). The action is now where the grass is; and grass, despite its tropical origin, is most successful in temperate climates.

Grass, of course, is widely distributed in the world, and with it herbivorous animals. But not all grasses are suitable for domestication, nor are all herbivores; and the species that are suitable are very unequally distributed. Eurasia is far better supplied on both counts than the rest of the world's landmasses; wild cows, for example, were found from Europe to China. And within Eurasia the most privileged region was the Near East; in particular, it was unusually rich in large-grained cereals. So it is no surprise that the Neolithic revolution began in the Near East—more specifically in the Fertile Crescent, the arc of cultivable land stretching from Palestine in the west to lower Mesopotamia in the east.

Just how the transition took place is not so easy to say. It helps to know that in and around Palestine there existed toward 10,000 B.C. a population that was already to a considerable extent sedentary—enough so for the huts they lived in to provide favorable environments for mice. What is more, one way in which this population survived was by harvesting the seeds of wild grasses, as we know from the hardware they left behind them—sickles, mortars, clay-lined storage pits, and the like. Assuming that these foragers had the same tendency to spill things as the rest of us, they would soon have found that they had inadvertently planted stands of grass in the vicinity of their huts. What they had done inadvertently they could go on to do deliberately. Meanwhile, they would be subjecting the grass to selective pressures that, if not quite natural, need not yet have been purposive—we will come to a good example of this in the next section. By now we are well on the road to domestication. We could continue in this vein, but the message is already clear. We can tell a plausible story of the emergence of farming with a certain confidence, but this is not quite the same as knowing how it actually happened.

After these early Near Eastern beginnings, farming both deepened on its home ground and made its appearance elsewhere.

Let us start with the appearance of farming outside the Near East. Before we consider how this in fact came about, it will pay to ask how it might have done so. First, we could imagine a world in which the emergence of farming was a very unlikely accident that just happened to occur once and once only. In such a world any subsequent appearance of farming in another region would arise from the spread of the original package. This spread might take either of two forms: a farming population could arrive and displace the local hunter-gatherers; or the local hunter-gatherers could themselves adopt the practice from farmers with whom they were in contact. Second, we could imagine a contrasting world in which farming emerged at the drop of a hat. In such a world, once the Holocene window had opened, farming would be likely to appear independently in numerous places, without any process of colonization or cultural borrowing to link them. So which was it?

As any reader inclined to compromise will suspect, it was a bit of both. The single most impressive and well-attested instance of the spread of farming from a center of origin involves the standard Near Eastern package of cultivated plants and animals. For example, within a few thousand years this package had made its way as far east as the northwest corner of India (by 5000 B.C.) and as far west as Britain (by 4000 B.C.). The only question we need ask is whether the practice was disseminated by colonization or by local adoption. Let us take the case of the westward spread of farming to Europe, since this is by far the best-known part of the story. For a generation it was thought, on the basis of what seemed to be a cogent correlation of archaeological and genetic evidence, that the central process here was one of colonization (an exception was made for southwestern Europe, particularly for the ancestors of the Basques). But recently more sophisticated analysis of both the mitochondrial DNA and the Y chromosomes of Europeans has shown that they are predominantly the heirs of the Palaeolithic population of Europe, with only a limited input from the Near Eastern population in Neolithic times. This is perhaps surprising: numerous hunter-gatherer populations the world over have fared far less well when farmers have appeared on their horizons, if only because of the tendency of farmers to outbreed them. But borrowing other people's culture is something humans are good at (though the earliest attested example of this process must be credited to the Neanderthals).

At the other end of the spectrum, an incontrovertible instance of the independent emergence of farming is provided by the New World—the Americas, as opposed to the Old World of Eurasia and Africa. Although the New World may have lagged several thousand years behind the Near East in this respect, developing farming in the fourth millennium B.C. rather than the ninth, there is nothing to indicate that it was subject to influence from the Old World. None of the American domesticated plants (notably maize) or animals (notably llamas) owed anything to the Old World, so the standard Old World farming packages played no part here. That leaves us to consider the possibility that someone who was aware of Old World farming might have tried to reproduce it in the New World by means of native species (a process known as stimulus diffusion). But this is extremely unlikely. There is no evidence of any kind of contact with Old World farmers in the relevant period, and at least in the case of maize, the process of domestication took so long that it could not plausibly have been the result of an attempt at imitation. So this instance establishes the category; we could go on to add some further examples where the case for independent emergence is strong.

What, then, comes out of this, other than a victory for moderation? First, the fact that we have at least two, and probably more, examples of the independent emergence of farming is instructive.

It confirms that there was something special about the Holocene, something that could produce convergent results in populations isolated from one another since the Pleistocene. But second, it is striking that the inhabitants of Britain—for example—should have waited several thousand years for farming to arrive from the Near East, rather than developing it for themselves on the basis of local species. So even in the Holocene, an independent emergence of farming must rank as a somewhat unusual event, albeit not so unusual as to be unique.

The Deepening of the Neolithic Revolution

Let us now go back to the Near East and take up what I referred to as the deepening of the Neolithic revolution. This term should not be taken to imply that the emergence of farming was an unqualified success. In fact, by about 6000 B.C. farming communities in Palestine were confronting an ecological catastrophe of their own making (they had burned far too much timber in order to manufacture lime to plaster the walls of their houses); and down to the present day farming has contributed greatly to the erosion of the Near Eastern landscape. But there is another side to the story, and for our purposes it is much more important. The oldest walls at the early Neolithic site of Jericho in Palestine date from around 8000 B.C. They belong to well-constructed round houses, and represent a dense permanent settlement inhabited by at least a couple of thousand people. This is intensive, not extensive, exploitation of an environment; and a more intensive way of life means more people and more culture. The very nature of the site, a hill created not by nature but by successive human occupations, provides a metaphor for the cultural accumulation that this lifestyle made possible, with each generation building on what the preceding generation had left behind. So once the platform of the original farming package was in place, numerous innovations could follow. A nice example is the plow, which appears in the fourth millennium B.C.: it presupposes prior development of domesticated animals to pull it and of domesticated plants to sow in its wake. To space out some other examples over the millennia, we could cite the making of pottery, the domestication of the olive, and the invention of the wheel; of these we will take up pottery in the last section of this chapter. But if we are to pick a single complex of innovations for a closer look at this point, it would have to be metalworking.

Stone tools have their limits. Figure 6 shows both sides of a flint knife from Predynastic Egypt (around 3250 B.C.), alongside two Palaeolithic hand axes (perhaps over a million years old). As you can see from the sections shown to the left of the knife, it has been worked with great skill to make it remarkably thin; this precise and regular implement shows none of the grossness of the Palaeolithic hand axes, whose thickness is apparent from the side views to the right. But this elegance carries a cost: the knife is fragile, and has in fact broken (two small pieces have gone missing as a result). Such a fate scarcely threatened the Palaeolithic hand axes, though to reach us intact they had to survive for a far longer period. There is indeed reason to believe that knives like this one were never intended for practical use: typically they are found in graves, rather than in the ruins of settlements.

Metalworking is the solution to this problem. A good metal can be worked thin and still be strong. Copper on its own is a bit too soft, but mixed with some tin or arsenic to make bronze it is not at all bad. Iron is much better, and steel better still. There are, however, a couple of difficulties to be overcome along the way.

One is that serious metalworking involves smelting ore and shaping metal at high temperatures—you cannot simply light a fire and go to work. Fortunately there is a cultural platform for this development. If you make pottery, you need to fire it, and (within the confines of a Neolithic culture) the higher the temperature at which you fire it, the better. If you also have building skills, you should be able to construct kilns in which you can fire your pot-

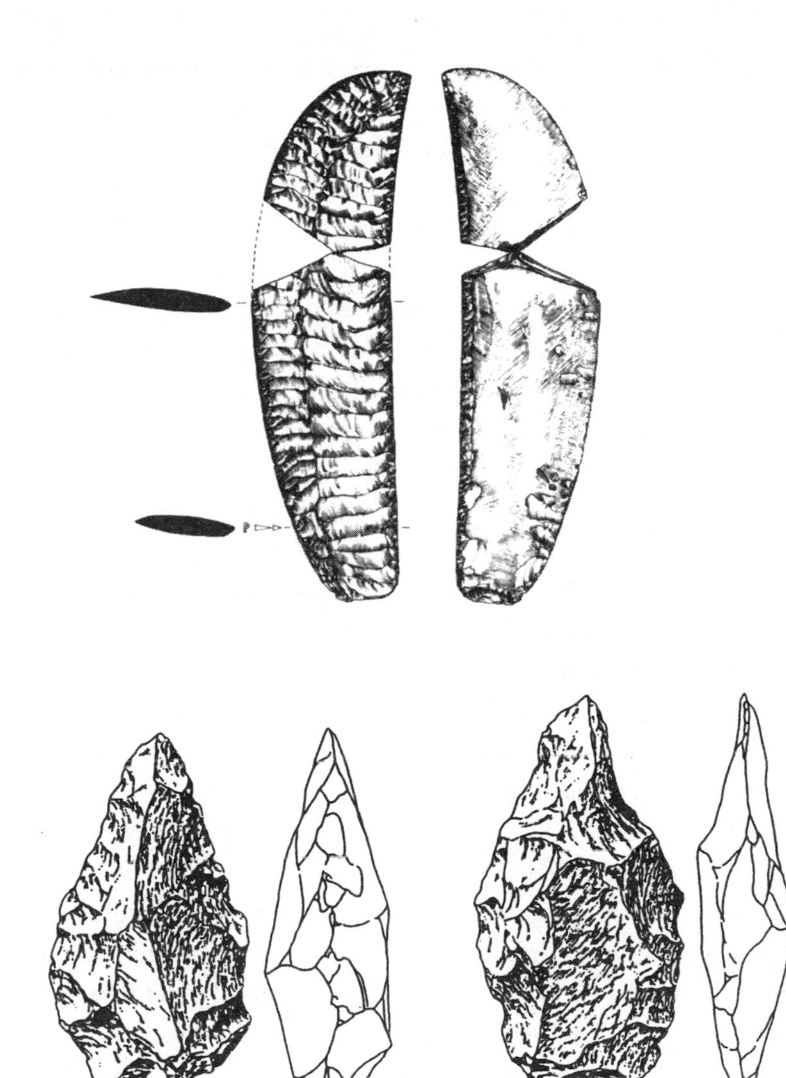

Fig. 6: A Predynastic Egyptian flint knife (*top*) with two Palaeolithic hand axes (*bottom*).

tery at suitably high temperatures. If you do this, the capacity you now have puts you within hailing distance of metalworking.

The other difficulty relates specifically to bronze. Copper, like other ingredients of bronze, is not in plentiful supply, and its distribution over the surface of the planet is uneven. This constrains the uses to which most societies can put it. Typically a Bronze Age society, like that of China or Greece in the last centuries of the second millennium B.C., is dominated by an aristocracy that uses bronze for its weapons and its ceremonial activities, while the mass of cultivators at the bottom of the society are still effectively in the Stone Age. The solution to this problem is iron. For reasons to do with the relative abundance of the elements produced in supernovas, iron is plentiful all over the earth; if you can reach the temperatures needed to smelt and work it, you can use it to make metal implements rather cheaply. It is thus with the Iron Age that we first encounter the widespread use of metal tools to cultivate the soil. At the same time the adoption of cheap iron weapons tends to destabilize traditional Bronze Age aristocracies, and so to open the door to new forms of social and political organization—a couple of which we will be looking at in later chapters. Here again, one thing is leading to another.

As we might expect from its pioneer role in the emergence of farming, the Near East was where the earliest developments in the history of metalworking took place. There was significant use of copper there by 5000 B.C. By about 3000 B.C. bronze was well established, with the ensuing Bronze Age lasting until about 1200 B.C. Thereafter ironworking became widespread, and for many purposes iron replaced bronze; in a sense the Iron Age has lasted ever since.

Like farming, metalworking later appeared in numerous regions outside the Near East. Thus in Britain and China, bronze was already being worked in the second millennium B.C., and iron in the fifth or sixth century B.C. These dates show a noteworthy acceleration: bronze working spread faster than farming, and ironworking spread faster than bronze working. They also give rise to the usual arguments as to whether or not such developments were independent. But there seems to be less evidence of independent development than in the case of farming; of the examples just named, only the emergence of Chinese bronze working is a serious candidate for independence. This should not surprise us: the more rapidly an innovation spreads, the less time or need people have to come up with it for themselves. The Americas are a different story. In the Andean region copper was worked in the first millennium B.C., and bronze from the first millennium A.D.; eventually metalworking techniques from this region also spread to western Mesoamerica. But iron metallurgy was unknown. This should not surprise us either: farming seems to have developed far later in the New World than in the Old.

We can now sum up our answer to the question at the head of this section. Just as the climatic mutability of the Pleistocene made it a good time for cultural agility, so also the climatic stability of the Holocene made it a good time for cultural accumulation. It was not that Neolithic cultures were long-lasting: typically their life spans are measured in centuries, in contrast to the millennia of Upper Palaeolithic cultures. But the basic innovations we have noted in this section were conserved from one culture to another, with no need for anyone to waste time reinventing the wheel. What made it possible for humans to take advantage of the Holocene opportunity in this way, apart from their native predisposition to culture, was that early in the Holocene they had come up with the greatest cultural platform of all time. It is because farming has played this role that we can unblushingly refer to its emergence as the Neolithic revolution.

II. The Genetics of Domesticated Plants and Animals

We saw in the preceding chapter how genetics can shed light on the origins of the human race. It can do the same for domesticated plants and animals, and in this respect it is now proving a valuable supplement to archaeology. In particular, it can provide crucial evidence as to whether a given plant or animal was domesticated once and once only, or at a variety of times and places.

One case where genetics establishes a unique domestication is a variety of wheat known as einkorn. Domesticated einkorn was already being cultivated in what is now southeastern Turkey in the ninth millennium B.C. A comparison of the DNA of numerous lines of domesticated and wild einkorn shows two things. The first is that domesticated einkorn is monophyletic—in other words, that all lines of the domesticated form go back to a common ancestor that sets them apart from the wild forms. This is enough to establish that einkorn was domesticated only once. The second thing shown by the genetic study of einkorn is that the domesticated form is closest to the wild lines found in a specific region to the west of the city of Diyarbakir. This gives us a bonus: we now have a good idea where einkorn was domesticated. Barley is a similar case. Here too it turns out that domesticated barley is monophyletic (though the analysis is complicated by the tendency of cultivated barley to cross with local wild forms). And here again identifying the closest form of wild barley helps to locate the domestication, though less precisely—this time in the region of Palestine.

With these monophyletic grains we can contrast what genetic studies have to tell us about the origins of domesticated cattle. They are not monophyletic. Examination of the mitochondrial DNA of Near Eastern cattle (the kind most of us are used to) and Indian cattle (the kind with the hump) shows that they have been diverging for well over a hundred thousand years—long before cattle were domesticated. In other words, the two species are the result of separate domestications from the wild. This need not mean that the two developments were historically independent: someone familiar with domesticated Near Eastern cattle could have domesticated Indian cattle in a process of stimulus diffusion. But the case is quite different from that of einkorn or barley. Genetics can also tell us something about the background of African cattle. In terms of their mitochondrial DNA—that is to say, in the female line—they are broadly similar to Near Eastern cattle; yet many of them are humped like Indian cattle, and it seems that in terms of their Y chromosomes—in the paternal line—they often owe more to India than to the Near East. In other words, they result from sexually asymmetric crossbreeding, doubtless brought about by their human owners. But was the female heritage of African cattle derived from cattle domesticated in the Near East? In the case of European cattle, the genetic evidence speaks clearly for a Near Eastern origin, and not for a separate domestication; yet African cattle are sufficiently distinct that they could well be the product of such an event. Again, stimulus diffusion would be hard to rule out.

Genetics can thus help to sharpen our picture of what happened in the emergence of farming. But there is another aspect of genetics that is worth attention here.

Domesticated plants and animals live in symbiosis with humans. Being humans, we tend to think of this relationship from the human point of view, but in biological terms it is a two-way street. Consider the evolutionary achievement of domesticated wheat. It has struck up a relationship in which it can rely on its human partner to render it a remarkable set of services: preparing the soil for it, planting it, protecting it from being eaten by other animals, freeing it from the competition of rival plants, harvesting it, and preserving its seeds through the winter. Of course, the cultivators take their cut—wheat is not a parasite—and this cut enables them to enjoy a far larger food supply than is available to their hunter-gatherer kin. But in return, domesticated wheat gets to grow the world over, and in quantities far beyond anything that could have been expected for its wild progenitor. To be sure, if humans were to go extinct, domesticated wheat would be in serious trouble; it has put all its eggs in the human basket. To date, however, it has done extraordinarily well for itself. Similar stories could be told about domesticated animals: at whatever cost to quality of life, they have become far more numerous than their wild relatives.

If we find this perspective teasing, it is not because it is in any simple way wrong. Since Upper Palaeolithic times, humans have increasingly been making waves; other species have thus had to come to terms with the human phenomenon. One obvious strategy was to get out of the way; hence the fear of humans so widespread among wild animals in most of the world (at least until they colonized suburbia). Another strategy was to get close to humans in such a way as to enjoy their protection; this is what successful domesticates did. Looking at the relationship this way does, however, bring out its asymmetry. One aspect of this is the central position of humans in the whole complex: for each domesticate the key relationship is with humans, not with some other member of the set. Likewise asymmetric is the manner in which the two sides have adapted to the symbiosis: for humans the adaptation is primarily cultural, whereas for wheat or cows it is not. This, after all, is why we can speak of the place of the cow in the way humans have made history, but hardly of the place of the human in the way cows have made history.

For plants and animals alike, the most important form of adaptation to domestication is genetic. Let us leave animals aside as too complicated, and take a simple example that concerns grain. In wild wheat and barley a process of "shattering" causes the seeds to fall to the ground as they ripen, thus impeding effective harvesting by humans (figure 7, *left*). In domesticated forms, by contrast, mutations in one or two genes are enough to solve the problem to mutual satisfaction (figure 7, *right*).

Fig. 7: Wild and domesticated einkorn wheat.

For humans it was different: as we already mentioned, the human adaptation to domesticates was mostly cultural, not genetic. Had it been otherwise, there would have been two kinds of humans: those who were genetically adapted to farming and those who were not. And this manifestly has not been the case. It was through being clever that humans adapted to those plants and animals with which they struck up relationships: it is clever to harvest grain, clever to dig storage pits, clever to sow seed at the right time of year, and so forth. In this respect they undoubtedly had a certain edge on the plants and even the animals, and this helps to explain how humans could adapt so rapidly to so many species at once, and why farming was to lead to so many further innovations. But the scope of this cleverness should not be exaggerated. No human could have grasped the emergence of farming as a process until it had already happened; on that level the humans involved had no more of a grand strategy than the plants or the animals. For example, it is unlikely that humans consciously selected lines of grain that did not shatter; if they simply harvested what they could with their sickles and planted from what they harvested, the effect would be to favor nonshattering mutants. The human contribution to the emergence of farming was doubtless the product of numerous such petty strategies along the way.

Yet in a couple of respects the human adaptation was markedly genetic. One is the ability to digest milk in adult life; this is widespread in populations with access to the milk of domesticated animals, but much less common elsewhere. The other, and more momentous, concerns disease. Old World populations, the only ones living in intimate symbiosis with large numbers of domesticated animals, had to come to terms with the germs these animals carried. Such germs have been at the root of infectious diseases that have killed many people in the Old World. But with the passing of the millennia, the populations of Eurasia and Africa built up a measure of genetic resistance. The effect was precisely to create two kinds of human, those who had this protection and those who did not. When the two kinds came into contact after Columbus sailed across the Atlantic, the effect of Old World germs on genetically unadapted populations was to be devastating. Domesticated plants, by contrast, have taken no such revenge on their human partners.

III. Pottery

Pottery is to the Neolithic what stone tools are to the Palaeolithic. Figure 8 shows some potsherds picked up at a site in the Near East in the summer of 2000. They are of varying dates; some are handmade, some were made on a potter's wheel. They are hardly

Fig. 8: Potsherds of various dates from an early site in the Near East.

museum pieces. But they are typical of the kind of thing that can be found on the surface of just about any site that has been occupied on a permanent basis by significant numbers of humans, provided the topsoil is not covered by modern buildings or lush vegetation. All of these sherds, of course, are fragments, and most of the breaks are ancient. This makes sense: a pot is typically used until it breaks, which it does sooner or later; after that the fragments are discarded, since few societies have yet thought of ways to recycle broken pottery in any quantity. Unbroken pots are mainly found in graves, almost the only context in which a pot goes out of circulation before it is broken. It is also worth noting that none of these sherds shows any sign of decoration, whether by impressing a pattern into the clay or painting it some color. This too makes sense: the average pot is a utilitarian object, and fancy decoration tends to be the exception rather than the rule. In contrast to these sherds, the pots illustrated in this book will indeed be museum pieces—sometimes beautiful, always instructive, but never typical.

As was already indicated, we have in pottery a phenomenon in some ways analogous to the stone tools we looked at in the preceding chapter. Clay, once it has been fired, is a good survivor (though not as tough as stone); recollect the curious clay objects illustrated in figure 1. Pottery thus has a salience similar to that of stone in the archaeological record, one that likewise exaggerates its role in the life of the people who made it. But this role, like that of stone tools, was still a significant one. For example, pots make it possible to heat water and cook soups or stews—a new lease on life for the toothless. They also provide mouse-proof storage; mice, be it remembered, are animals that succeeded in living off us for thousands of years without giving us a cut in return.

But unlike stone tools, pottery is a phenomenon of the Holocene, and more specifically of farming societies. Palaeolithic hunter-gatherers, in so far as they were nomadic, would have had little use for pots: they are heavy and fragile, exactly the kind of thing one does not want to carry around. This means, among other things, that pottery was not part of the common Palaeolithic heritage of Asia and the Americas. Like farming, it must have developed independently in the two regions.

And yet the broad fit between pottery and farming is not perfect. Not all farming societies have had pottery. Some, like the earliest Neolithic societies in the Near East, had not yet invented it; its absence there is so striking by later standards that archaeologists speak of the Pre-Pottery Neolithic. Other societies had it at some point but lost it; the Maori of New Zealand stemmed from a culture that had once possessed pottery, but did not preserve the skill in the course of migrations from one Pacific island to another.

More interestingly, a few Holocene hunter-gatherer cultures did have pottery. Japan provides the leading example. The phase of Japanese prehistory known as the Jōmon period stretches from around the tenth millennium to the first millennium B.C.; the oldest known pots, from the island of Kyūshū, may date from the eleventh millennium B.C. (Even older dates have been claimed, and would take us back to the tail end of the Pleistocene.) By about 7500 B.C. finds of pottery are plentiful. Jōmon pottery, however, was coarse, thick, and low-fired—its makers probably used nothing more advanced than a bonfire (quite unlike the potters of Neolithic China, for example). This pottery nevertheless suggests what can be done in a hunter-gatherer culture rich enough to be sedentary—one resource available in unusual abundance in this case being fish. The real puzzle is perhaps why more such privileged hunter-gatherers did not develop pottery. One early European culture of the Upper Palaeolithic produced large numbers of objects made of fired clay; yet we do not have a single Magdalenian pot.

All this is intriguing, but it should not be allowed to get too much in the way. The broad association of pottery with farming is telling us something important: that farming makes it a lot easier to accumulate more material culture than you can carry. That in turn helps us to see, in very concrete terms, why the emergence of farming was a revolution.

CHAPTER 3

THE EMERGENCE OF CIVILIZATION

I. Did Humans Make the Only Kind of History They Could?

So far in this book we have asked simple questions and fared reasonably well in finding answers. It may be rash to depart from this prudent course, but looming in front of us is a more complicated question that we will have to face sooner or later. Take the broad outline of history as it has played out in the last ten thousand years. This history was obviously one possible outcome of the combination of unusual climatic conditions and the capacities of behaviorally modern humans. But was it the only possible outcome? Or did an accident, or a chapter of them, determine that history was to be the way it was, and not some other way?

The first thing we should admit is that, had we surveyed the earth from some lofty eminence on the eve of the Holocene, the chances that we would have correctly predicted the course of history—even in its broadest outlines—are virtually nil. One fundamental reason for this is that we are dealing with *emergent* phenomena, the kind that regularly appear when entities we understand pretty well interact at a new level of scale and complexity. "More is different," as one physicist has put it. So even if we had understood the way humans lived in the Upper Palaeolithic far better than we now do, this does not mean that we would have been able to say what sort of society would emerge in the event that it became possible to have many more people and much more culture.

Our goal, however, is far less ambitious, since we are content to pose our question in retrospect. What we want to know is whether more could have turned out to be different in ways quite unlike the actual course of history. Yet even this sounds like an invitation to science fiction—how else could we set about imagining histories incommensurable with what we know, and speculating as to whether they were possible or probable outcomes of human history? Ideally, what we would like to have is, of course, a controlled experiment. We would go back to the late Pleistocene, divide humans into two separate populations, deny them any subsequent contact with each other, and return in ten thousand years or so to check the results. We would then be able to see whether the histories they had made in mutual isolation were quite different, or just local variations on the same basic pattern. Would two kinds of history have emerged, or just one?

The Old World and the New

It is our peculiar good fortune as students of the past that nature set up this experiment for us, and did so without our having to take responsibility for the consequences (it was not, as it turned out, in the least an ethical experiment). Sometime in the late Pleistocene —just how late is controversial—humans from the Old World occupied the New World for the first time by crossing what is now the Bering Strait, but was then a land bridge. The experimental conditions were not perfect. The New World was a significantly different environment from the Old, in ways that we can take up in chapter 5; and the seal between the two worlds was imperfect even before it was dramatically broken in 1492. But it was as good an experiment as we are going to get, and the results are well worth our attention.

One way to approach these results is through the eyes of an early Spanish observer of Mexican society, the "Anonymous Conquistador." In the accounts they have left us, conquistadores are typically concerned to recount their glorious deeds. The Anonymous Conquistador is an exception: he has nothing to say about himself and his companions, and instead gives us a sober, matter-of-fact description of early sixteenth-century Mexican society. What sort of a society is it?

Like the Old World society from which our conquistador stemmed, the New World society he describes is based on farming. For example, he speaks of the grain from which the people make their bread (clearly maize). This is accordingly a sedentary society, indeed one with large cities. These cities have streets and squares that he compares favorably to those of his own country; the main square of Tenochtitlán is about three times the size of that of Salamanca. They hold markets at regular intervals, with an orderly arrangement of wares; their most commonly used currency is the cacao bean (he provides a rate of exchange). Society is stratified, with lords at the top. Unlike the people at large, the lords dine sumptuously. The men hold women in lower esteem than do any other people on earth; they are polygamous, like the Moors (i.e., Muslims). Organized political power is a salient feature of the society: at the summit there is a ruler resembling an emperor, and there are kings and such (not to mention some cities with non-monarchic governments). War is waged by armies, which are divided into companies, with officers in command of them; the Aztec ruler has a special guard of ten thousand warriors. Weapons include bows and arrows, spears, swords, and slings. Religion plays a prominent part in the life of the society, which our conquistador describes as very devout. It is associated with special buildings (he calls them temples or mosques), special personnel (he compares them to bishops and canons), idols or gods, and rituals that he compares to Christian ones (one ritual reminds him of matins). The sons of lords are trained in the temples. In short, the basic categories that our conquistador brought with him from the Old World (Christian and Muslim) seem to have worked well enough for him, even though in the New World he was confronted with a civilization that had evolved quite separately from his own.

Was there an aspect of Mexican society so utterly unlike anything known to him that he could not make head or tail of it? He seems to have had no difficulty describing polytheism, cannibalism, and human sacrifice, features of the society not paralleled in early modern Spain, though attested elsewhere in the Old World. If there was an aspect of Mexican society that was totally opaque to him, he does not mention it. Of course, he may well have misunderstood things. And some of his cross-cultural comparisons, though helpful for his intended audience, would make modern academics squeamish. Yet it would be pointless to quarrel with the broad outlines of his account: Mexican society was indeed sedentary, stratified, and endowed with cities, markets, kings, armies, religion, and the like.

The most striking differences between the two societies in fact have to do with things that were commonplace in the Old World but largely absent from the New. Our conquistador tends not to remark on these absences, but we can easily supply them. Mexican farming (in contrast to that of the Andes) lacked pastoralism. Mexican technology lacked a whole string of things. For example, it made no use of the wheel in transport or in pottery making, and its metalworking was limited in character and impact. This in turn explains why lithic technology was so highly developed. Our conquistador, who was greatly impressed by this, remarks on the stone spearheads and wooden swords with inset stone blades; Old World societies had little need for such things. All this can broadly be attributed to the less favored environment of the New World, which meant that things happened later there; as we saw in the preceding chapter, metalworking fared better in the Andean region than in Mesoamerica, but even there it took the Spanish conquest to initiate an iron age.

Our conclusion from nature's experiment is thus fairly clear-cut. The millennia during which the Old and New Worlds were cut off from one another had seen two major processes unfold in each: the emergence of farming, followed by that of civilization. The trajectories of the two worlds were admittedly far from identical. Indeed, their respective forms of farming did not have a single domesticated species in common; and the two civilizations differed in numerous ways. All this difference is serious enough to require us to look at the major regions of the premodern world one by one in the second part of this book. Yet this should not make us lose sight of the basic unity. What initially emerged in both the Old World and the New was farming, and not two radically different ways of life; in the same way, what later emerged in both was civilization, and not two things so divergent as to send us in search of a new concept.

So the answer to our initial question seems to be yes: there was only one kind of history waiting to emerge from the coincidence of behaviorally modern humans and the Holocene. As we will see, it could emerge sooner, or later, or not at all, depending on environmental conditions and interactions with other populations. But where it did come onstage, it was recognizably one thing, and nothing suggests that something completely different was lurking in the wings. This is why you still do not know whether the stone walls in figure 2 belong to the Old World or the New.

The Oldest Civilizations

So much for our initial question. The purpose of the rest of this section is more prosaic. I have dropped the term "civilization" into the argument as though it were as transparent as "farming." In fact it is not. In part this is because it is traditionally a term of approbation, and therefore subject to the pulling and tugging that affects all desirable labels. But it is also, and more seriously, because it is vague; like so many words in natural languages, it is getting at something important, something we need to get at, but not in a precise kind of way. To remedy this, at least in part, I will do two things. I will run over what are commonly seen as the major instances of the more or less independent emergence of civilization (we need not concern ourselves here with the spread of already existing civilizations). And I will try to pick out a couple of the features of these emergent societies that make us want to identify them as civilizations.

Unsurprisingly, our list opens in the region of the Near East, where two civilizations emerged around 3000 B.C. The first was probably that of Mesopotamia or, more specifically, of the Sumerians (the earliest people we can call after a name they used themselves). The second, in the northeast corner of Africa, was that of Egypt. From the Near East we move to northwestern India, where the Indus Valley civilization makes its appearance in the middle of the third millennium B.C. The second millennium B.C. then takes us east to China and west to Crete. Finally, the first millennium B.C. brings us to Mesoamerica, in the first instance to the culture we know as Olmec. This list could perhaps be shortened by disallowing some of these civilizations on the grounds that they were influenced by earlier ones, and it could certainly be lengthened by applying the term "civilization" more generously. But for our purposes this list is good enough. One thing worth noting about it is its geographical unevenness. In terms of continents, we have one instance from Africa, one from Europe, one from the Americas, and no fewer than three from Asia. In another way it is even more uneven: only China and Mesoamerica take us outside the Near East and the regions adjoining it.

What, then, makes us want to include these instances on our list? At a very general level, complexity is the name of the game. When we looked at Mexico through the eyes of the Anonymous

Conquistador, we saw a society that was complex in the same kind of way as his own: it was sufficiently large and specialized that, within it, different kinds of people did different kinds of things in a systematic and organized way. We thus have no problem classifying highly complex societies as civilizations, to the exclusion of thoroughly simple ones. The question is what to do with the ones in between. We could just draw an arbitrary line, or we could specify a bundle of properties and require that a civilization have most of them. But the term "civilization" suggests that there is in fact some kind of a quantum leap in complexity along the way, a kind of change of state, and not just a smooth, continuous transition. This is certainly something that many of those who study the societies on our list tend to believe.

Assuming that they are right, what would we pick as the outward and visible signs that this leap has taken place? The following sections will take up two obvious candidates: the development of writing, and the appearance of highly developed kingship.

II. Writing

We take it for granted that writing is so sophisticated a cultural practice that it could emerge only in a complex society; and the historical record undoubtedly tends to bear us out. But why should this be so? What was there to prevent earlier farming societies, and even hunter-gatherers, from developing writing?

A functional writing system is made up of two very different components. We can refer to them as the hardware and the software.

The hardware needs of a writing system are obvious: we must have something to write *on*, and something to write *with*. These days we normally write by making marks on paper with ink. The ink is not a serious problem; cave painters in the Upper Palaeolithic used pigments that would serve the purpose. The paper is the product of a more advanced technology, one that developed in

China around the turn of our era and only gradually spread to the rest of the world. Before paper became universal, there were great regional variations in what people wrote on: clay in Mesopotamia, papyrus in Egypt, bamboo in China, bark in Mesoamerica, and so forth. Some materials, like papyrus and bark, required elaborate preparation, but others, like clay and bamboo, would have been easy enough to use in the Upper Palaeolithic. And even without them, there was nothing to stop the artists of the time from adding written captions to their cave paintings. The problem, then, is not in the hardware.

On the software side, writing requires a system for representing language—in other words, a way of transposing something we hear into something we see. This is easier said than done.

Today, over most of the world, people do it by means of the alphabet, in one or other of its innumerable variants. But however much we take the alphabet for granted, there is good reason to think that developing it was not at all easy. There is a strong case for the view that the world's alphabets are—to borrow the term from genetics—monophyletic. In most cases a shared origin is clear from careful comparison of one script with another. Thus the alphabetic script you are now reading is not an English invention; it derives from that of the Romans, who got it from the Etruscans, who had it from the Greeks, who borrowed it from the Phoenicians, who lived in the region where it was originally developed at some time in the second millennium B.C. In a minority of cases, the forms of an alphabet show no convincing resemblance to other scripts, but the historical context nevertheless makes stimulus diffusion the most likely explanation (meaning that someone got the idea of writing from someone else, but then devised a script of his own). This is likely to be the origin of the main family of Indian scripts (dating from the fourth century B.C., or somewhat earlier), and it is unquestionably behind those of the Georgians and Armenians (dating from the fourth or fifth century A.D.). In short, the alphabet is the product of a development that took place only once in the Old World, and never in the New. To this we should add that it took well over a millennium for a region already familiar with writing to come up with the alphabet. Clearly there is something very sophisticated about the phonetic analysis of spoken language on which an alphabet depends.

The earliest writing systems are not, of course, alphabetic. Typically they are hybrids; some signs stand directly for words, while others stand for sounds (say, a syllable). This duality is likely to reflect the way the scripts developed, though the only case in which we are well informed about this is the cuneiform script of Mesopotamia, which evolved in the later fourth millennium B.C. and was probably the earliest in the world. Here the record shows two major stages leading up to a full-fledged writing system. In the first, we find clay tokens that seem to stand for animals, commodities, and numbers. This is a system of representation, but what is represented is not language but the things themselves. In the second stage, these signs come to be tied to particular words with particular sounds; thus they can also be used phonetically, or supplemented with purely phonetic signs (though this does not happen in the case of early Chinese writing). Eventually this process takes the system to the point at which there is not just a correct way to understand the signs but also a correct way to read them. At that point we have a real writing system, which will typically comprise several hundred signs.

Developing this software requires a lot of mental effort, but unlike metalworking, it has no particular technological prerequisites. Why, then, couldn't people have made this effort long before the Sumerians? The answer is that they perfectly well could have, and maybe even did; brilliant and eccentric people can hardly have been a monopoly of the last five thousand years. The question is whether such a development could have caught on. Early writing is a technology that imposes heavy costs—there has to be a community of people who have learned those several hundred signs. Two things in particular point to the burdensome character of the writing system in early Mesopotamia. First, writing was in itself a profession, the defining activity of scribes (whereas today it is a prerequisite for just about any profession). Second, the scribes themselves did not find writing second nature: alongside their tablets recording useful information about sheep and grain, they were also producing large numbers of tablets that served simply as reference lists of signs.

The reason why writing took so long to emerge should now be clear. It was not the inherent difficulty of procuring the hardware or developing the software that stood in the way. Rather, it was the need for an appropriate social structure. Somebody had to have a strong need for this information technology, and a willingness to pay handsomely for it by maintaining a community of otherwise unproductive scribes. Such a need and such a willingness are hallmarks of a complex society. To put it crudely, early writing presupposes a powerful state—which over most of human history has meant some form of kingship.

But before we go on to kingship, we should return for a minute to the alphabet. Even this writing system is not cost-free. Nature adequately motivates our children to learn to talk, but they learn to write in the coercive setting of school; and illiteracy is a serious problem in many parts of the world even today—far more so than aphasia. Yet, without any question, the alphabet is much easier on the learner than cuneiform. One historically significant result of this is that, with the rise of the alphabet, writing ceased to be so closely tied to professional scribes and complex societies. Northern Arabia in pre-Islamic times was too arid and poor to rank as a complex society, but its rocks are covered with graffiti in alphabetic scripts.

III. Kingship

Bird Jaguar, who stands on the right in figure 9, is every inch a king. His captive kneels haplessly before him, bound with a rope, and

Fig. 9: Bird Jaguar attending to a captive.

manifests his submission by touching his right shoulder with his left hand. For the moment he has been left alive, but he is likely enough to undergo ritual sacrifice in due course. A subordinate lord stands discreetly on the left, leaving the king as the undisputed center of attention.

As New World civilization goes, this representation is not particularly old. Bird Jaguar ruled the Mayan city of Yaxchilán in the eighth century A.D. But it is easy to be reminded of a much older version of the same theme from halfway around the globe, the Narmer Palette (figure 10). Narmer was the ruler of Egypt around 3000 B.C. Here, on the left, he holds his captive by the hair (as Mesoamerican rulers sometimes do), about to smite him with a mace; it would have been pointless for him to keep his victim alive, since the Egyptians did not sacrifice their captives. To the left of the king is the humble figure of his sandal bearer, and at the bottom are the corpses of two further enemies of the king. Each of

Fig. 10: The Narmer Palette.

these figures is identified by a caption. By contrast, the ten decapitated corpses on the other side of the palette (shown on the right) do not rate such treatment; quantity takes over from quality. The scale here is much smaller, but again the king towers over everyone else—including the corpses, the attendant who carries his sandals, and the diminutive standard-bearers.

Did Bird Jaguar really capture the roped figure, or was the king simply taking the credit, as one Mesoamericanist has suggested? Did some wretched captive really spend the last seconds of his life waiting for Narmer to smite him with his mace? An Egyptologist has argued that Narmer was in fact no great conqueror, and that what we see here is not a historical event at all. We have no way to recover the truth, any more than we have access to the political jokes that doubtless circulated at the time. But if the truth is hopelessly opaque, the message comes through loud and clear. A true king is bigger and better than everyone else. Unlike a hero, who fights at least some of his antagonists on equal terms, the king is the inevitable victor of every battle. He strikes terror into the hearts of his enemies, and sooner or later he slaughters them.

If the king's enemies are in deadly peril, his loyal followers and his obedient people should be safe and sound. In a poem of about 1800 B.C., the chief treasurer of the king of Egypt admonished his children in this sense. On the one hand, he spoke of the king's rage, and its dire consequences; a rebel gets no tomb, and his body is cast into the water. On the other hand, he stressed the king's mercy; he gives food to those who are in his service, and he makes the land green, filling it with strength and life. Such talk would have been perfectly intelligible in early Mesopotamia. There too there is no lack of the iron fist—"the mace that controls the people." But we also hear rather more of the velvet glove. Rulers are described as shepherds, an exploitative role but not a predatory one. They deliver prosperity, milking the udder of heaven. Hammurapi, ruler of Babylon in the eighteenth century B.C., explains in the prologue to his famous legal code how he was called upon "to improve the living conditions of the people" and "make justice appear in the land." This royal idiom also stresses popular acclaim. In the twenty-fourth century B.C., Lugal-zagesi tells us how the land rejoiced at his rule, and spells this out city by city. Such emphasis on public relations may reflect the political structure of early Mesopotamia: it was made up of numerous city-states, and the local assembly of a city might even choose a king. Mesopotamian kings, like those of Egypt, were bigger and better than their subjects, but they perhaps felt more need to underline the services they rendered in order to justify the difference.

One way in which kings never quite succeeded in being different was that sooner or later they died like everyone else. "Well! Would you believe it?" a Frankish king exclaimed petulantly on his deathbed in A.D. 561. "What manner of king can be in charge in heaven, if he is prepared to finish off great monarchs like me in this fashion?" Yet something kings could and did do in the face of this celestial insult to their dignity was expend enormous resources in an attempt to ensure that even in death they were maintained in the style to which they were accustomed. The Great Pyramids of the twenty-sixth century B.C. remain to this day imposing monuments to this quest and its massive social cost.

But there was another, more sinister aspect of the royal way of death that came and went long before Egyptian kings took to building pyramids. King Aḥa, who ruled in the thirtieth century B.C., was the first king to be buried on a grand scale, and he did not embark on his journey to the next world alone. Those he took with him, presumably his followers, included considerable numbers of young men. The next king, Djer, was buried with nearly six hundred persons. This was not an Egyptian idiosyncrasy. A few centuries later, burials of members of the royal family at Ur in Mesopotamia included numerous men and women, not to speak of animals. In the late second millennium, the same practice is well attested in China under the Shang dynasty. In each of these cultures, funeral customs subsequently softened and the practice disappeared, though the ruthless emperor who unified China was still accompanied in death by numerous concubines in 210 B.C. In the New World the practice had not died out at the time the Spanish arrived in the early sixteenth century. When the Tarascan ruler of Michoacán in western Mexico died, he took with him over forty men and women to serve him, including seven lords, some doctors, a storyteller, and a jester. After getting drunk, they were beaten to death with cudgels. The practice attests a conception of kingship guaranteed to send a shiver down any nonroyal spine; nothing suggests that the emergence of civilization was good for human rights.

We should not think of dramatic inequality as an innovation of the last few thousand years. There is a Palaeolithic burial in Russia dating from more than 22,000 years ago in which two children were interred in clothing so elaborate that it included some ten thousand beads; these alone would have taken thousands of hours to make. Yet the potential for inequality in human societies must have increased greatly with the emergence of farming and the developments that followed, notably those that issued in the earliest civilizations. Elaborate kingship, then, can broadly be seen as a product of an increasing social complexity, a development that is independently documented in the archaeological record of the societies we have been looking at. There is, however, no way to tell just when and where humans first used words that we could validly translate as "king."

Whatever the antiquity of its origins, kingship was to prove remarkably durable. In the last few thousand years competition between polities has been intense, and it has been rare for a line of kings to last more than a few centuries. But until the French Revolution, kingship (with occasional episodes of queenship) remained the normal form of government for complex societies: the fall of one king or dynasty issued sooner or later in the rise of another. At the heart of the institution there are often, perhaps always, two incompatible conceptions of its nature: one in which kings exist for their subjects, and another in which subjects exist for their kings. The bones of the followers who accompanied early kings in death offer a vivid testimony to the reality of the second.

PART TWO

THE SMALLER CONTINENTS

CHAPTER 4

AUSTRALIA

I. A Continent of Hunter-Gatherers

Where the World's Land Is

The Holocene was (and fortunately still is) a global phenomenon. So it makes sense to see the human response to it in global terms, and this is what we have been doing so far. We started with the Upper Palaeolithic, the period in which behaviorally modern humans spread over all the world's continents except Antarctica. We then went on to the emergence of farming on all the inhabited continents with the exception of Australia. Finally we turned to the appearance of civilization on the same set of continents. We could try to continue in this vein, but there would be a long wait before the next comparable milestone: the process whereby the world's populations have been integrated—or, more bluntly, thrown together—into something approaching a single global society. When that process began is hard to say; in this book I have chosen to start it with the Islamic expansion, followed in due course by the European expansion, and eventually by the emergence of the modern world. But between these initial and final global stories there lies a mass of history that does not fit well into such a framework. Here it makes better sense to get acquainted with the major cultural regions of the world one by one, rather than to encounter them as fleeting examples decorating a global narrative. So we will now begin to make our way slowly through the smaller continents, starting with Australia, after which we will move region by region through Eurasia.

But before we set aside our global perspective to concentrate on Australia, we should take this opportunity to examine the general layout of the world's landmasses. (We focus on the land, not the sea—for all their intelligence, sea mammals have left the making of history to land mammals.) Unlike climate, the lay of the land (see map 1) is a reliably stable feature of the backdrop to history. Continents drift, but they do so sufficiently slowly that they have been in their present locations for a good many million years, and will remain there for quite some time to come. The only significant exception to this is—from the point of view of plate tectonics—cosmetic. Sea level depends on the amount of ice stacked at or around the poles; the result is that the world's continental shelves are more extensively exposed during ice ages, and less so during warm spells like the Holocene.

Most of this is very familiar, but to have a sense of how different the arrangement of the continents might be from what it is today, we have only to glance back a quarter of a billion years. That takes us to one of the times when all the world's landmasses were assembled in a single supercontinent, in this instance Pangea. But this unity was eventually disturbed. One disruptive event was the separation of Pangea into two parts divided by a mid-world ocean: Laurasia lay to the north, Gondwana to the south. It is the asymmetrical fates of the two that interest us.

All things considered, Laurasia has held together rather well. Eurasia remains by far the largest landmass in the world, and it has been augmented by two significant fragments of Gondwana: India has moved north to collide with Asia and thereby create the Himalayas, while Arabia and the Fertile Crescent have more or less deserted Africa for Asia. North America, however, has moved away from Europe, to such an extent that it is now attached to Asia by a land bridge that stands above sea level during ice ages. To the north these landmasses encircle the polar region, but they do not occupy it.

The unity of Gondwana, by contrast, has been shattered. Apart from the desertions just mentioned, one continent occupies the South Pole, where it plays a vital role in setting the world up for ice ages, and preventing it from being much hotter than it is; but this location rules out any part for Antarctica in human history. That leaves us with three continents, none of them comparable in size to Eurasia: Africa, South America, and Australia. Plate tectonics has placed them in positions in which they are widely separated from each other. Instead of forming a coherent unit, the southern continents are subtended, so to speak, from the landmasses of the Northern Hemisphere. Thus Africa, though separated from Europe by what remains of the mid-world ocean (except on occasions when it dries up), is almost joined to Europe in the west, and actually joined to Asia in the east. South America has been linked to North America by a land bridge for the last few million years. Australia, the smallest inhabited continent, is more isolated. In an ice age, it forms a single landmass with New Guinea and Tasmania, while much of island Southeast Asia becomes part of the Eurasian mainland; but even then, there is no land bridge between them.

What comes out of such considerations is a set of asymmetries. Comparing north and south, we find a lot more land in the north, and it is much more consolidated. Comparing east and west, we find a lot more land in the Old World than in the New. Comparing the Eurasian landmass with the other continents, we see that it is not only the largest but also unusual in that its axis runs from east to west, not from north to south. Each of these asymmetries played a part in shaping a world history in which the north set the terms for the south, the Old World for the New, and Eurasia for the Old World. But fortunately for the diversity of history, these effects have become dominant only in the last few centuries. That leaves us with a long period in which human cultures were much more diverse than is the case today—or presumably was when the human race first emerged. The length of this period owes much to the relative dispersal of the continents in the current geological epoch, and the plurality of more or less separate habitats they provided. A human history unfolding on Pangea would have been a very different story.

The Absence of Farming in Australia

What would the world we actually live in have looked like if the Neolithic revolution had not happened—if it still lay in the future, or was not going to happen at all? The question is idle, but thanks to the dispersal of the continents we can put it to work on a smaller scale. In several parts of the world there were regions where hunter-gatherers prevailed until modern times, as in large parts of the Americas. More helpfully still, Australia provides us with an experiment in which a few hundred thousand hunter-gatherers had a continent to themselves until the eighteenth century. What did they make of it?

We should first ask why there was no farming in Australia prior to the eighteenth century. One reason is obvious: the isolation of the continent greatly reduced the chances of the arrival of farming from outside. This isolation goes very deep. Greater Australia—the combination of Australia, New Guinea, and Tasmania—has been separated from any other continent bar Antarctica for some seventy million years. Since parting from Antarctica it has moved closer to Southeast Asia, but, as we already noted, the two landmasses remain separated by open sea even in an ice age. As a result, few placental land mammals made the crossing before modern times; bats, rodents, humans, and dogs were the exceptions. This in turn made it possible for Australia to retain numerous species of nonplacental mammals, notably marsupials, as late as the time when humans first arrived. South America, by contrast, had lost most of its marsupials several million years earlier, when the land bridge was formed that now links it to North America.

In recent times there have been contacts between parts of northern Australia and the islands farther north. There were relations between the tribes of the Cape York Peninsula and the island peoples of the Torres Strait; but these are unlikely to be more than a few thousand years old, and their impact was geographically limited. Farther west, there were annual expeditions from Macassar (on Sulawesi) to Arnhem Land, with the purpose of collecting sea slugs for the Chinese market; but this activity seems to be even more recent, perhaps no older than the eighteenth century. What marked the real end of Australian isolation was not these marginal contacts in the tropical north but rather the arrival of the Europeans in the temperate south. This began with the voyages of the Dutch navigators of the seventeenth century. Then, in 1788, the British established their first settlement at Sydney in southeastern Australia. Thereafter two centuries of colonization drastically reduced the size of the native population, and effectively put an end to the hunter-gatherer way of life that isolation had protected for so long. Another of nature's unethical experiments had come to an end.

A second reason why farming did not appear in Australia before modern times is that the continent provided a distinctly unfriendly environment for its independent emergence. Rather flat and heavily eroded, Australia is short on recent geological disturbance. There are highlands along the east coast, but the only really impressive mountains are in New Guinea, which was cut off from

Australia by rising sea levels around 7000 B.C. This geological passivity has had the effect of limiting soil formation, which is one reason why Australia is not particularly fertile. Another reason is that much of it is arid, not to say desert, particularly in the center and west (see map 2); and even in the best-watered regions the rainfall is unreliable. These conditions imposed significant limits on Australia's plants and animals. There was forest, particularly on the east coast where the highlands brought down the rain. But grasslands were not extensive. Moreover, the larger nonplacental mammals, many of them herbivorous marsupials, suffered extinction after the arrival of humans, and very likely because of it. Altogether, there was not much to domesticate.

These conditions do not mean that farming could not possibly have emerged independently in Australia—as it may have done among a closely related population in the quite different environment of highland New Guinea. They do mean that such an emergence was a distinctly unlikely event. There is therefore nothing surprising about the fact that it did not happen before the Europeans introduced their own package of domesticated plants and animals in modern times.

An Outline of Australian Prehistory

The story of humans in Australia until the eighteenth century was accordingly a story of hunter-gatherers, and by the same token a story without metalworking, cities, kingdoms, or written records. Nor did the Australians make pottery. This means that, for most of its course, we can know about Australian prehistory only from the same kinds of sources as we possess for the European Upper Palaeolithic: an abundance of stone tools, eked out with some skeletal remains, some rock art, and the like. We can sum up the outlines of the resulting picture in a few words.

It is well established that there were humans in Australia around forty thousand years ago. Dates that might take this back another twenty thousand years have not been confirmed, and even older

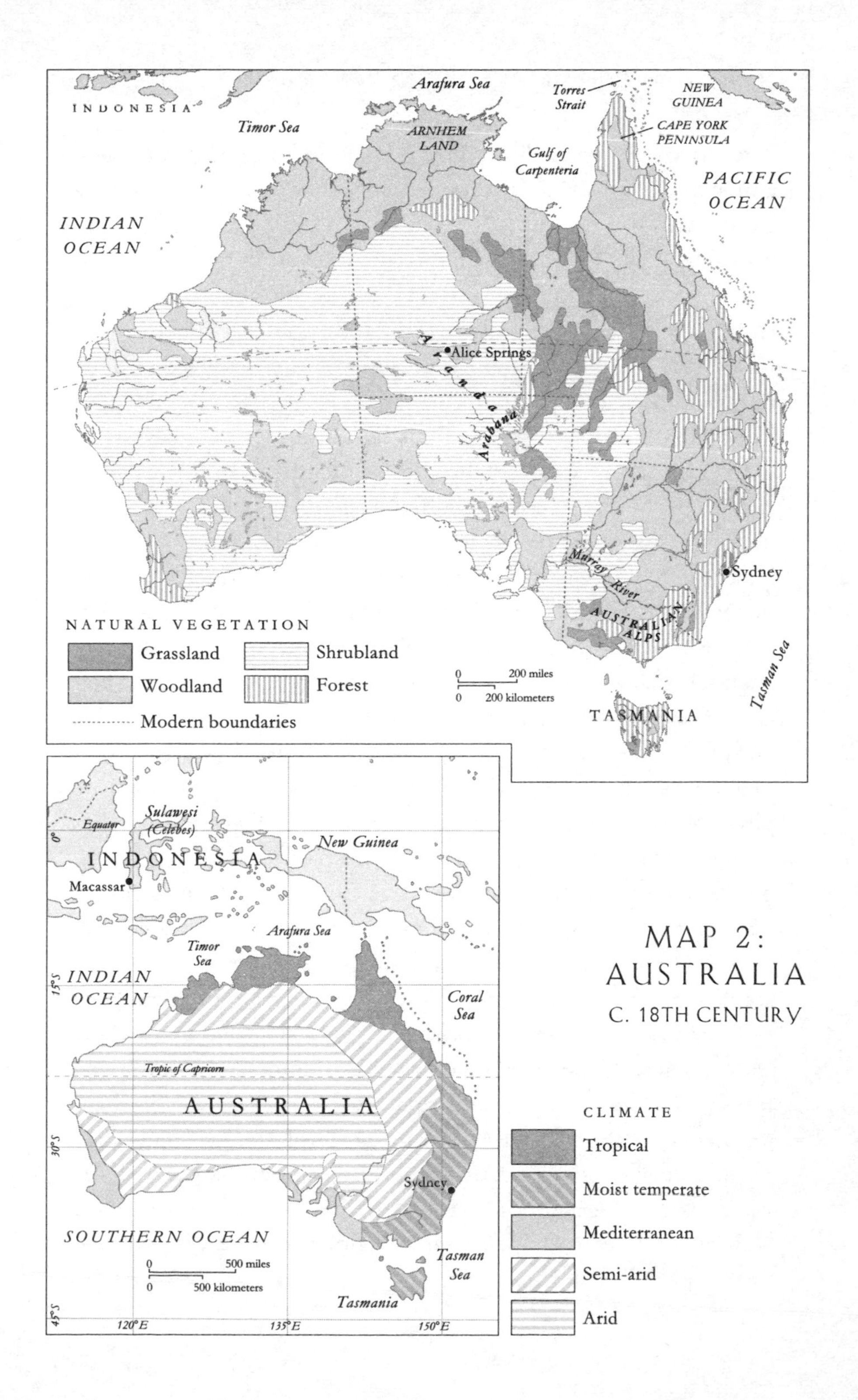

INDONESIA
Arafura Sea
Torres Strait
NEW GUINEA
Timor Sea
ARNHEM LAND
CAPE YORK PENINSULA
Gulf of Carpenteria
PACIFIC OCEAN
INDIAN OCEAN
Aranda
Alice Springs
Arabana
Murray River
Sydney
AUSTRALIAN ALPS
NATURAL VEGETATION
Grassland
Shrubland
Woodland
Forest
Modern boundaries
0 200 miles
0 200 kilometers
TASMANIA
Tasman Sea
Equator
Sulawesi (Celebes)
INDONESIA
New Guinea
Macassar
Arafura Sea
Timor Sea
INDIAN OCEAN
Coral Sea
Tropic of Capricorn
AUSTRALIA
Sydney
SOUTHERN OCEAN
0 500 miles
0 500 kilometers
Tasman Sea
Tasmania
0°
15°S
30°S
45°S
120°E
135°E
150°E
MAP 2: AUSTRALIA
C. 18TH CENTURY
CLIMATE
Tropical
Moist temperate
Mediterranean
Semi-arid
Arid

dates have been discredited. This gives modern humans the same antiquity in Australia as in Europe. Since they had to cross the sea from Southeast Asia to reach Greater Australia, they must have had some kind of raft or boat; this is borne out by the fact that in the same early period they were also able to colonize islands to the east of New Guinea. By thirty thousand years ago humans were widespread in Australia, though they may not have moved into the more arid parts of the interior until Holocene times. As in Upper Palaeolithic Europe, there are finds here and there to make us sit up: we encounter some of the world's oldest beads, for instance, and probably some very old hafted axes (roughly axes with handles—one example, which is around thirty thousand years old, looks pretty much like the kind still in use in the nineteenth century). We are particularly lucky that some wooden implements dating back about twelve thousand years survive in a swamp, including the first known boomerangs. But to paint a rounded picture of these early Australians is beyond us.

For the last few thousand years the archaeological record is much richer, though just what it adds up to is not easy to say. More plentiful evidence does not necessarily imply that anything has changed: artifacts dating from the recent past naturally tend to be better represented because they have not had to survive for so long. But it does seem that recent millennia, though still a continuation of the hunter-gatherer way of life established in Australia some forty thousand years ago, were significantly different from earlier times. The issues are somewhat perplexing, and we can best leave them until we take a closer look at some stone tools that reflect the change; we will do this in the last section of this chapter. In the meantime let us take advantage of the fact that when farming finally arrived in Australia, it brought ethnography in its wake. Archaeology, after all, is a somewhat dismal science; unlike archaeologists, ethnographers can describe for us what it is like for the people they study to be alive.

II. Getting Married among the Aranda

In the late nineteenth century the Aranda were one of the largest tribes (or ethnic groups, if you prefer) of central Australia. They numbered at least two thousand people spread over a territory extending for a few hundred miles in the arid region around Alice Springs. They were divided into local territorial groups that might camp together, though more often groups of one or two families would wander around on their own. On a typical day the women would be out gathering grass seeds, or digging for lizards and honey ants; the men might hunt kangaroo or emu, or stay in the camp and sleep. They had no houses or tents, but would make use of a lean-to as a windbreak. They wore virtually nothing. The main part of a woman's equipment might consist of a digging stick, a wooden trough for carrying things, and a pair of grindstones. The men would have spears, spear-throwers (to one end of which a sharp piece of flint or quartzite would be attached), shields, hafted stone axes and knives, and boomerangs. Such essentials apart, the Aranda did not carry much around with them, not did they store things other than certain sacred objects associated with totem groups. This inventory of material culture is instructive. In the normal course of things, the most likely elements of Aranda culture to survive the centuries and be recognizable to later archaeologists would have been the stone tools. Even the name of the tribe would eventually have been lost. Instead, the Europeans arrived. This was disastrous for the Aranda; but we are fortunate that their traditional culture was rather carefully studied and recorded before it disintegrated.

As this account makes clear, Aranda society was in material terms a simple one, so we might expect to find the same simplicity in its nonmaterial culture. And to a considerable extent we do. Political authority, for example, was almost nonexistent. A man might be accorded the respect of others because of his skill in hunting or fighting, or his knowledge of tribal traditions. Such a man would exercise influence, particularly if he also held the ritual office of leader of a totem group. But there were no chiefs. Likewise, leading men of different local groups might get together to resolve problems and make decisions, but there was no authority extending to the tribe as a whole. Another instance of cultural simplicity, and a startling one, is the Aranda number system (not that it was anything unusual in Australia). There were only two numbers with names of their own: "one" and "two." Thereafter the Aranda would use a form of binary arithmetic to count a little further: three was "two-and-one," four was "two-and-two," and five was occasionally "two-and-two-and-one." Beyond that they spoke only of "many."

In such a society, one might expect the rules governing marriage to be comparably simple. All that would really be needed would be some kind of rule that had the effect of preventing too much inbreeding. A simple way to do this would have been to require people to marry out of their totem groups. According to tribal mythology, these totem groups originated at the beginning of things when certain beings made it their business to turn a variety of animals and plants into humans; in each case the resulting group would be called after the animal or plant from which it had been metamorphosed—the emu people, the witchetty grub people, and so forth. There were groups of emu people, for example, scattered in different locations in the tribal territory, each such group having its ritual leader and sacred objects, and forming the core of a local territorial group. Totem groups were widespread in other tribes, and rules requiring people to marry out of them were common there. But among the Aranda the totem group played no part in the rules governing marriage. There was a rule of comparable effect that pushed people to find spouses in different localities and families, but it had nothing to do with the totem groups.

The main rules governing marriage were of a quite different

kind. Their starting point was the division of the tribe into two halves—two moieties, as the anthropologists term them. These moieties were a significant feature of the social structure; for example, they affected who sided with whom in a conflict, and the layout of large camps (which was determined by complicated rules). Moieties were common in Australian tribes, and they could be expected to have names; but in the case of the Aranda, this was not so. Taking a cue from their binary arithmetic, let us designate the two moieties as 0 and 1. In its simplest form, the marriage rule required people to marry out of their moieties and assigned the children of a marriage to the moiety of the father. Though it may seem a little complicated to include both totem groups and moieties in the same social structure, the marriage rule as we have stated it is simple enough to grasp.

But this was much too simple for the Aranda, who went on to divide each moiety into two sections. These sections had names, but let us continue with our binary notation: moiety 0 was divided into sections 00 and 01, moiety 1 into 10 and 11. Reformulated to take the sections into account, the marriage rule now required that a person in a given section of one moiety marry into the corresponding section of the other: 00 into 10, 01 into 11, and vice versa. A child was placed in the same moiety as the father, but in the other section of it; for example, if the father was in section 00, the child was in section 01.

According to tribal myth, the marriage rules were devised by the emu people at some early stage in the history of the tribe. Four local groups of them were involved, two in the south and two in the north. The leader of one of the southern groups proposed the system just set out, and this was adopted by the southern groups. But the leaders of the northern groups, who were unusually wise, condemned the plan and instead proposed a more complex one calculated to make things "go straight." The four leaders met, and it was decided to adopt this plan. A grand tribal initiation ceremony was held, and at the end of it all the people stood up and, where necessary, women were redistributed in accordance with the new rules. Nothing, incidentally, suggests that women played any part in making these decisions.

What the new system did was divide each section into two subsections; so there were eight subsections in all, although the Aranda could not have said this. The northern Aranda gave names to the subsections, whereas the southern Aranda did not—a fact that is doubtless connected with the roles of the northern and southern emu groups in the myth. But how did the subsections work? If you can bear it, let us extend our binary notation to a third digit: subsection 001 is subsection 1 of section 0 of moiety 0. In terms of this notation, the marriage rule now says that you marry into the corresponding subsection of the corresponding section of the opposite moiety: if you are in 000, you marry into 100, if you are in 001, you marry into 101, and so forth. But what about your children? It now depends which moiety you are in. As before, let us assume you are the father. If you are in moiety 1, then your child is in the *corresponding* subsection of the other section of the same moiety (so if you are in 100, your child is in 110). But if you are in moiety 0, then your child is in the *other* subsection of the other section of the same moiety (so if you are in 000, your child is in 011). If you are clearheaded and mathematically gifted, you will have no trouble figuring out how all this works from the woman's point of view; remember that the rules are unisex with regard to marriage itself, but not with regard to the allocation of the children.

There is one final nicety about these rules. Other tribes had different systems, and this could give rise to problems. For example, a man of another tribe might come to live among the Aranda, bringing his wife or seeking one there; or a woman might be captured from another tribe, and assigned to an Aranda man as a wife. These tribes, too, had their marriage rules, but different ones. What was needed if things were to go straight was a set of conventions for relating one set of rules to another.

On the south, for example, the Aranda were neighbors of the Arabana. Among the Arabana, getting married seems to have been simpler than it was among the Aranda. There were two named moieties, and all you had to do was to marry into the other moiety; in contrast to the Aranda rules, the Arabana rules placed the children in the moiety of the mother. There were no sections, let alone subsections. So how could the Aranda bring the two systems into a harmonious relationship? Mercifully, the solution as we have it refers only to moieties and sections. Let us say that an Arabana man has arrived among the Aranda, with or without a wife; the old men then decide to which section he is to be assigned. Say that he is assigned to section 01; his wife (whether old or new) will then belong to 11, and their children to 00. Now for the clever bit. Whereas the relationship of sections to moieties is given for the Aranda (0 consists of 00 and 01, 1 of 10 and 11), it is arbitrary for the Arabana, provided the husband and wife fall into different moieties. We use this license to effect a transposition, assigning the Aranda section 00 to the Arabana moiety 1 (and likewise the Aranda section 10 to the Arabana moiety 0). Now go back to the case of the man in 01 with a wife in 11 and children in 00. If we construe this situation on the Aranda rules, the children are in the father's moiety; but if we construe it on the Arabana rules, they are in the mother's moiety. Everyone is satisfied.

We have no direct knowledge of the history of the Aranda system. Subsections dominated a large area in northern Australia, with the Aranda at its southern edge; and although many different languages were spoken in this region, the names of the subsections were surprisingly uniform. This suggests that we have to do with a system that had originated in one particular place and spread to others at a fairly recent date. There is in fact some reason to believe that subsections may have arisen in an interaction between two tribes, each of which had four sections. Such sectional systems are widespread in Australia. But when and how did *they* come into being? Through some merging of two earlier moiety systems? And if so, why in each case does the more elaborate system catch on so widely?

If this exposition of Aranda marriage rules leaves you baffled, be thankful that you live in what is in this respect a decidedly simpler society. Whether or not you understood the details, it should be obvious by now that in this aspect of their culture the Aranda were operating a system of considerable complexity. This was not the only such aspect; their ritual life was also remarkably elaborate, and they invested much time and energy in it. Clearly this complexity was compatible with the material circumstances of their life, and undoubtedly some features of it were beneficial to them in material terms. But there is no way we could adequately explain the form of their marriage rules on the basis of such circumstances alone (subsections existed in both the tropical north and the arid center of Australia, two very dissimilar environments). The Aranda could perfectly well have arranged these matters differently, as did other hunter-gatherers known from the ethnographic record. The Bushmen of southern Africa, for example, have no moieties, sections, or subsections. Who you can marry is likely to turn on a distinction between joking relationships (in which you can be cheeky and salacious) and avoidance relationships (in which you must be respectful and distant). Both systems have their root in complex algebras of kinship, but the branches are distinctly different.

What this shows is that hunter-gatherer societies possessed a potential for apparently superfluous cultural innovation and diversity that we can scarcely hope to recover from their material remains. As a conclusion about hunter-gatherers, this is far from trivial: the overwhelming majority of them lived in prehistoric times, and are thus known to us only by such remains. But there is also another way to look at this. Humans, it seems, have a quite remarkable capacity to tie themselves and others in knots by devising elaborate and ultimately arbitrary rules. Among the Aranda this tendency ran wild even in the absence of the most basic amenities of a farming society. This coexistence of cultural convolution and material simplicity reveals with extraordinary clarity one of the key constituents of human societies. We have to wonder what this gift for extravagant rule making would lead to when humans, or some of them, acquired a level of material resources sufficient to give society a plasticity it never had for the Aranda.

III Flaked Points and Other Novelties

Back to archaeology, and the question what has been going on in Australia over the last few thousand years. Figure 11 shows Australian stone tools of two different periods. Those on the left are edge-ground hatchet heads; they date back to the Pleistocene and are about eighteen to twenty thousand years old. Those on the right are flaked points for spears; they belong to the Holocene and date from the last millennia B.C.

Though the functions of the two types were different, it is hard not to see a significant contrast. Relative to one another, those on the left look primitive, while those on the right look advanced. Their level of craftsmanship is much finer, the objects are more delicate, and often they are smaller; nothing like them is attested for the Australian Pleistocene. These flaked points are by no means a flash in the pan. They are reckoned to appear about the fourth millennium B.C. and to begin to disappear in the first; they are strongly represented in the north and south, though not in the east and west. Moreover, to archaeologists they are just one element in a flurry of new types of stone implement that come, and sometimes go, at different times and places in Australia over the last few thousand years (though not in Tasmania, which was cut off from the mainland by rising sea levels about fourteen thousand years ago). For our purposes this one example is enough. How, then, are we to explain the proliferation of such novelties in the lithic record?

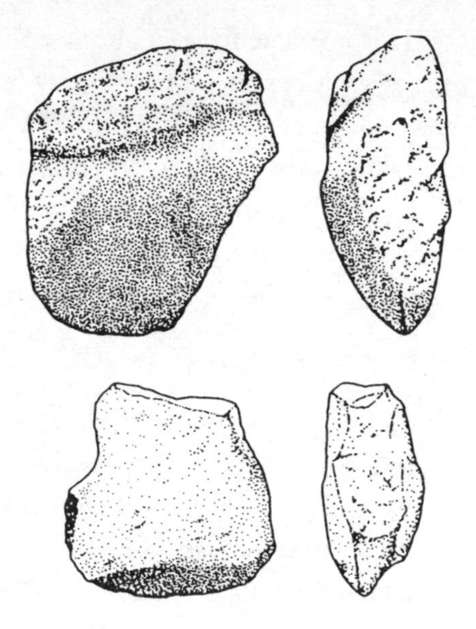

Fig. 11: Australian stone tools of the Pleistocene (*left*) and later Holocene (*right*).

Before we try, we should do what we can to place these lithic changes in the wider context of the last few thousand years of Australian prehistory.

Here archaeology has several further contributions. First, humans now appear in places where we do not meet them before: the "Australian Alps" in the southeastern highlands, for example, or the small islands off the Australian coast (in Tasmania, too, the human presence became more extensive). Second, there is reason to believe that human exploitation of the environment became more intensive. One example of this is the evidence of increased seed harvesting in the last two thousand years, a labor-intensive practice attested by the survival of grindstones; it has been argued that Aranda society as we know it could not have existed without this development. Another example is the large cemeteries that suggest the presence of unusually dense populations living off the resources of the Murray River in the southeastern interior (a den-

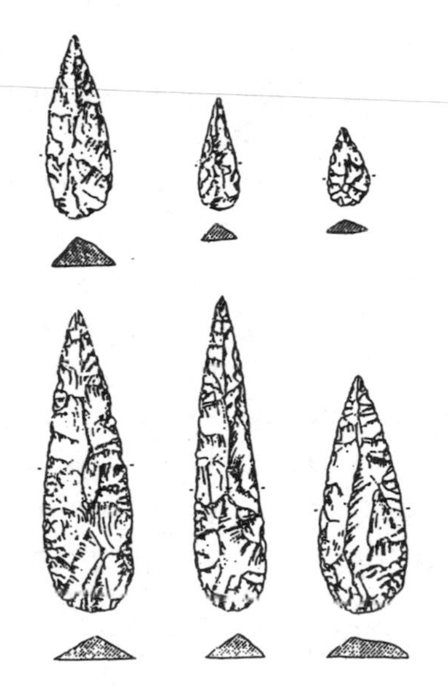

sity that in due course must have made these societies a sitting target for Old World germs). Third, a new mammal arrived from Southeast Asia. The dingo descends from the domesticated dog of Southeast Asia, itself of Indian origin. Once in Australia it ran wild, but some human groups adopted dingoes and made use of them in hunting. The dingo is first attested in Australia about 1700 B.C.; this is in the south, even though the animal must surely have entered the continent from the north. Once established, it seems to have added a couple of species to the list of marsupial extinctions on mainland Australia.

It is also possible that the mutual relationships of the languages of Australia have something to tell us about prehistory; this is another kind of evidence for the past that our Muslim historian Ṭabarī did not think to use. There were perhaps some 250 distinct languages in Australia when the Europeans arrived; today this number has been reduced by an order of magnitude, but something is known of many of the languages that have been lost. The linguists have so far failed to reduce them to any kind of family tree (in contrast, for example, to their success in showing that English, Dutch, and German descend from a single protolanguage, West Germanic). But Australian languages have shared features, like the absence of an "s" sound; and the vast majority of them have a lot in common. For reasons we need not go into, this majority is termed Pama-Nyungan; it covered the whole of mainland Australia except the northwest, where non-Pama-Nyungan languages were the rule. The question is what sort of process might have generated the common features of Pama-Nyungan.

One possibility is that the features in question originally belonged to a single language spoken at a particular place and time, and that the speakers of this language, or their culture, spread over most of Australia at the expense of older peoples or cultures. With the passage of time the original Pama-Nyungan language would then have broken up into a mass of local dialects, and these would gradually have diverged from each other. The significant point is that for the residue of common features still to be detectable in modern times, the spread and breakup of Pama-Nyungan would have to have taken place within the last few thousand years. This might tie in with the archaeological record.

The other possibility is that the common features of the Pama-Nyungan languages do not point to a common ancestry, but rather reflect the outcome of a process whereby neighboring languages rubbed off on one another. Archaeology shows us that material objects traveled long distances in prehistoric Australia, doubtless by being given or traded from tribe to tribe. Ethnography confirms this and adds culture: a dance, for example, could be transmitted over remarkable distances, just as seems to have happened with the subsection system. There was even a drug trade. With this kind of interaction, linguistic traits could easily have been passed along from tribe to tribe, and perhaps over a long period they could have come to characterize most languages of Australia. This too would be of interest, but for what it would tell us about space rather than time.

How do we put all this together? It depends on a number of things.

One is how far we want to see all the various phenomena as aspects of a single story. Here the least adventurous course would be to dismiss them as unrelated: to regard the dingo as an isolated import, the lithic technologies as unconnected even to each other, the more extensive and intensive exploitation of the environment as a matter of disparate local developments, and the common features of Pama-Nyungan as the result of a process of rubbing off. The most adventurous approach would be to see a single process at work. Even the development of more elaborate social structures such as we saw among the Aranda could be worked into the picture.

Another issue is how much weight we want to give to contact with Southeast Asia. There has to have been such contact, since the foreign origin of the dingo is beyond doubt; but it was manifestly not enough to bring about the introduction of farming. The key question is whether we should attribute the change in stone tools to such contact. We certainly can: an appropriate lithic technology was present in Southeast Asia in the relevant period. But we may prefer to see Australian prehistory as unfolding in relatively splendid isolation.

A final question comes down to our taste in causes. The easiest thing to get hold of here is the palpable novelty of the stone tools. So did this improved technology lie behind the new patterns of exploitation of the environment? And could Pama-Nyungan have spread in association with it? Yet, for all we know, the key change could have been located in some aspect of society or culture that is archaeologically invisible. If we want to play it safe, we should probably fall back on the view that many factors interacted; such a claim may not shed much light on anything, but at least it is hard to refute.

In the absence of compelling evidence, in this and many other such questions, the choice is yours. But two things emerge very clearly from the last few millennia of the Australian experiment. The first is that the hunter-gatherer condition does not preclude material and cultural change of a kind that looks suspiciously like progress. The second is that, without the appearance of farming, these changes will not lead on to cities, kingdoms, and the like.

CHAPTER 5

THE AMERICAS

1. From Alaska to the Tierra del Fuego

Their arrival in the Americas in the late Pleistocene brought humans into a world with a layout very different from that of Australia. Whereas Australia is an island, South America is permanently joined to North America in the present geological epoch, and North America in turn is joined to Asia during ice ages. Just as striking is the contrast in climatic range. Australia is confined to temperate and tropical bands within a single hemisphere; North America extends far into the Arctic, while the tip of South America comes within some hundreds of miles of Antarctica. But the advantages in intercontinental competition are not all to the Americas. Although the Americas are far richer than Australia in terms of the number of climatic bands they contain, none of these bands has a length comparable to those of Eurasia. This matters, because innovations—most obviously domesticated plants, but other things too—spread more easily within climatic bands than between them. Land may stretch continuously for some 8,500 miles from Alaska to the Tierra del Fuego; but from east to west, the dimensions of the Americas are far less impressive—around 3,000 miles where each continent is at its widest. This distinctive layout does not in itself tell us much about the habitat humans would find when they reached the Americas, but it does yield a first approximation. We can already start to wonder how human societies would fare in such an environment.

But before we introduce the human element, we should sketch in the geography of the Americas a little further (see map 3). Our primary concern is with the Holocene (before that conditions were rather different). Since the picture is very roughly symmetrical, let us begin in the middle. Here, just as in the Old World, we find a broad tropical band that includes large expanses of rainforest (though there would have been much less of it in the more arid conditions of the last ice age). The distribution of the tropical band between the two continents is, however, very unequal. South America gets the lion's share, since this continent extends on both sides of the equator and is at its widest in the tropics; whereas only the southernmost part of North America lies within the tropical band, and at this latitude the continent is at its narrowest.

North and south of the tropics, each continent has its temperate band; this may be forest, grassland, or desert, depending on the rainfall. In North America the temperate belt falls where the continent is at its widest, and we encounter substantial amounts of all three kinds of terrain: forest in the east, grassland in the middle, and desert in the southwest. The makeup of temperate South America is similar, but the continent is much narrower at these latitudes (though somewhat wider in an ice age).

Beyond the temperate bands lie the Arctic and Antarctic regions of the two continents. North America has a large amount of land as far north as the Arctic Circle, and this territory is similar in character to the Arctic regions of the Old World: as one goes north, a belt of Arctic forest—the taiga, in Old World parlance—gives way to a bare, open wilderness—the tundra. South America, by con-

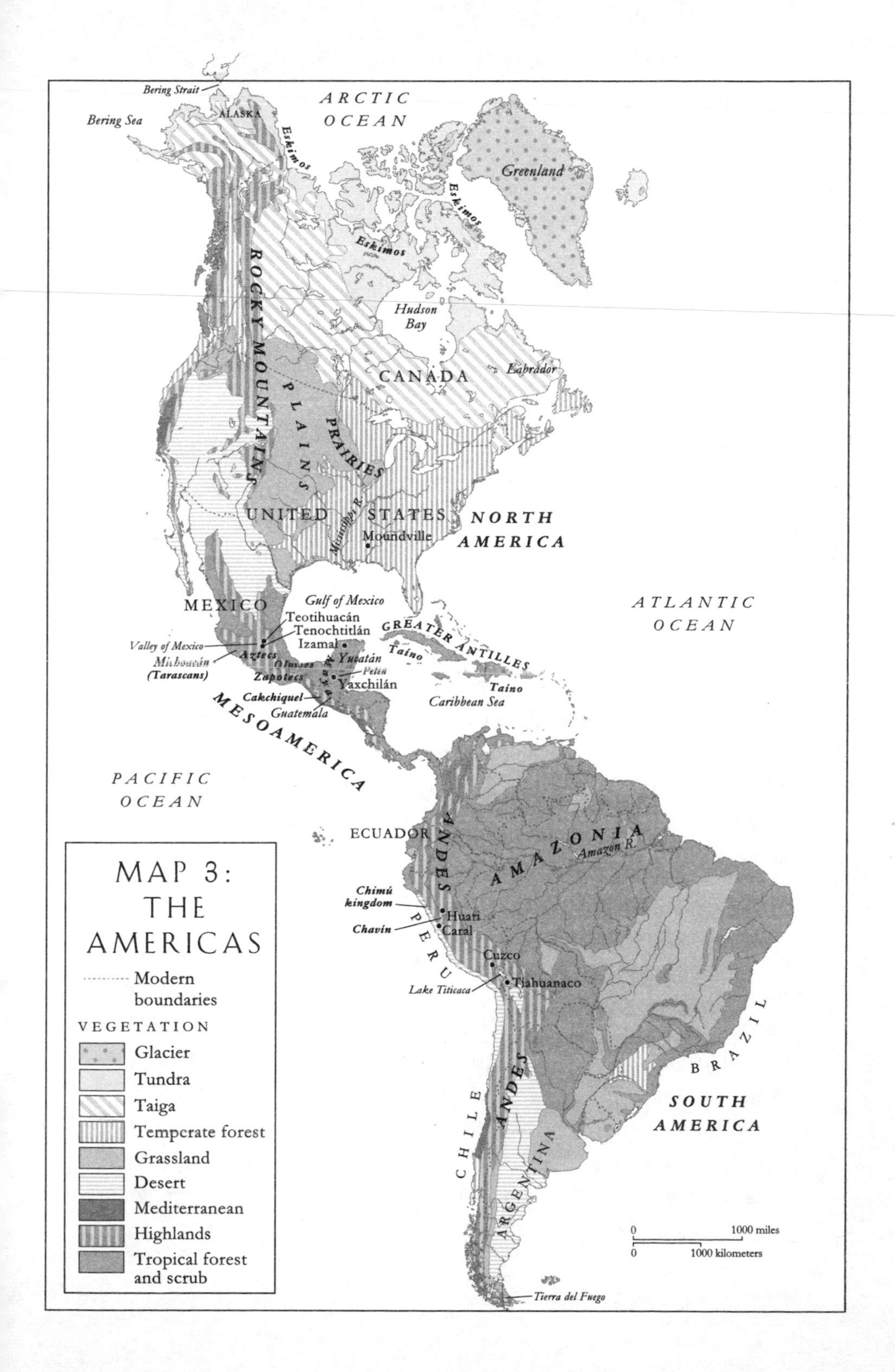
MAP 3: THE AMERICAS
Modern boundaries
VEGETATION
Glacier
Tundra
Taiga
Temperate forest
Grassland
Desert
Mediterranean
Highlands
Tropical forest and scrub
Bering Strait
Bering Sea
ALASKA
ARCTIC OCEAN
Eskimos
Greenland
Hudson Bay
CANADA
Labrador
ROCKY MOUNTAINS
PLAINS
PRAIRIES
UNITED STATES
Mississippi R.
Moundville
NORTH AMERICA
Gulf of Mexico
MEXICO
Teotihuacán
Tenochtitlán
Izamal
Valley of Mexico
Aztecs
Olmecs
Zapotecs
Maya
Yucatán
Petén
Yaxchilán
(Tarascans)
Cakchiquel
Guatemala
GREATER ANTILLES
Taíno
Caribbean Sea
ATLANTIC OCEAN
MESOAMERICA
PACIFIC OCEAN
ECUADOR
ANDES
AMAZONIA
Amazon R.
Chimú kingdom
Chavín
Huari
Caral
PERU
Cuzco
Lake Titicaca
Tiahuanaco
BRAZIL
SOUTH AMERICA
CHILE
ARGENTINA
0 1000 miles
0 1000 kilometers
Tierra del Fuego

trast, stops well short of the Antarctic Circle, and possesses only small areas of taiga and tundra.

The single most important thing missing from this sketch is the mountains. Unlike Australia, the Americas have their share of geologically recent mountain chains. These run from north to south, against the grain of the climatic bands, and are located along the western side of each continent (where the seafloor is being subducted under the edges of the American continental plates). In the north the highlands are broad but not particularly elevated; the best-known component of the system is the Rocky Mountains. In the south the Andes tend to be both narrower and higher—but where they form two parallel ranges with a plateau in between, they provide a kind of corridor running north and south. These highland areas confuse our simple picture of climatic bands, but they matter enormously.

The Peopling of the Americas

Like Australia, the New World was uninhabited by humans (or any species of ape) until the late Pleistocene. A glance at a map shows that they were most likely to enter the Americas from the far north, since any other point of entry would have required a willingness and capacity to cross oceans with few island steppingstones. To the northeast the distances between islands were less formidable than at lower latitudes, but still demanded a seafaring ability that appeared only with the Vikings (alias the Norsemen) in the late first millennium A.D.; and even they had no discernible impact on the mainland of North America. That leaves the northwest as the obvious gateway from Asia to the Americas, though a rather peculiar one. It had to be negotiated in two stages. First, the prospective immigrants had to wait for an ice age so that the Bering Strait would turn into a land bridge and give them access to Alaska. Then they had to wait until the ice age was over so that glaciers blocking their movement to the south should melt. This makes the end of the Pleistocene a plausible context for the human occupation of the Americas.

We have no direct indication that this is what happened, but the indirect evidence fits. With regard to the route, genetic testimony points strongly to affinities between the populations of the Americas and those of Asia, especially Siberia. The date is more of a problem. It is clear that northeast Asia was inhabited by 13,000 B.C. In Alaska the oldest known sites go back to 11,000 or 12,000 B.C. Farther south there is no dispute about the presence of humans by 11,000 B.C. But earlier dates from various parts of the Americas have their champions, and may yet prevail, though at this point they remain controversial. Of these earlier dates, that currently taken most seriously would place the appearance of humans in the southern cone of South America at about 12,500 B.C., implying a yet earlier date for their initial arrival in North America. But the difference, though highly significant for specialists, is hardly mind-blowing for the rest of us. Whatever consensus develops, there is no real doubt that the occupation of the Americas was a much later event than that of Australia.

So what did these relatively recent immigrants make of the Holocene window in the Americas?

American Hunter-Gathering

The first thing that stands out is the prominence of hunter-gatherers throughout the pre-Columbian period. In contrast to the Australians, they did not have either continent to themselves; but relatively speaking they continued to occupy far more territory than their counterparts in the Old World.

The northern half of North America offers an obvious example. At a latitude at which the peoples of northern Asia were tending domesticated reindeer, those of North America hunted caribou. Since caribou and reindeer are close enough to be considered the same species, the explanation of the difference is probably to be sought on the human side. Perhaps the Old World development was a case of stimulus diffusion: the peoples of the Eurasian Arctic were exposed to pastoralists tending herbivores in the grasslands to their south, whereas, for a reason we will come to, those of the American Arctic were not.

Farther to the south, much of what is now the western United States was occupied by hunter-gatherers, and many tribes of the plains and prairies, though practicing some agriculture, lived primarily by hunting and gathering. Temperate South America presents a similar picture on a smaller scale. Most of the land was left to hunter-gatherers; they hunted the guanaco, the wild ancestor of the llama, but did not domesticate it. Likewise, hunter-gatherers were to be found here and there in the tropics.

As in Australia, the persistence of the hunter-gatherer way of life did not mean an absence of change. In the far north, for example, the arrival of the Eskimos—or their "Palaeo-Eskimo" predecessors—from northeast Asia within the last few thousand years marks the appearance of a culture specially adapted to Arctic hunting. From Alaska they spread eastward to Greenland, and westward back into the northeast corner of Asia—thus becoming the only American population to colonize the Old World. They were likewise the only pre-Columbian people in whose tool kit iron had a place; they must have obtained it from meteorites and from trade with northeast Asia or the Vikings of Greenland. To the south of the Eskimos, we encounter two families of languages occupying vast territories, Athapascan in the west and Algonquian in the east; the existence of these incontrovertible families points to expansions within the last two or three thousand years, and it does so far more clearly than Pama-Nyungan in Australia. What powered these expansions we do not know. Turning to South America, we find pottery dating back to about 6000 B.C. in the eastern part of the tropical region; this was made by groups exploiting aquatic resources, but innocent of farming (rather as in Jōmon Japan).

Of course, not all hunter-gatherers of the Americas lived in, or took effective advantage of, environments that allowed such developments. The American Antarctic, for example, was no mirror image of the Arctic. Though neither as poor in resources nor as harsh in climate as the Arctic, the tip of South America was inhabited by societies as simple as any in the world; their canoes, for example, were far less seaworthy than those of the Eskimos. There was an obvious reason for the disparity: in contrast to the Arctic, this Antarctic world was a very small one, and completely isolated from similar territories on other continents. Yet overall, the American hunter-gatherer scene was as changeable as that of Australia.

Why did hunter-gatherers, however simple or complex their societies, continue to occupy so much of the Americas? The answer lies not in any distinctive strength of their way of life but rather in the relative weakness of American farming.

American Farming

Old World farming typically has two domesticated components: plants and animals. The first thing to note in the Americas is the much less prominent role played by animals. Such domesticated animals as there were tended to be small, like the turkey in Mesoamerica and the guinea pig in the Andes. The only domesticated herbivores were the Andean camelids, the llama and the alpaca, and they did not spread to other parts of the Americas. Overall, the problem was not lack of grass or, initially, of herbivores. But just as in Australia, the arrival of behaviorally modern humans was followed by massive extinctions among the larger animal species of both continents (the evidence for the role of humans being much more specific in the American than in the Australian case). The result was that, by the time farming developed, few suitable animals remained to domesticate; this left large areas of the American grasslands to the hunter-gatherers.

In the realm of domesticated plants, one major limitation relates to the techniques of cultivation. Whereas the farmers of the Old World used the plow, those of the New World depended on the digging stick—a difference obviously related to the lack of oxen in the Americas. This made New World agriculture a labor-intensive affair—"horticulture," as it is often called. It also meant that some potentially very fertile land could not be cultivated with the prevailing technology; again this land was left to the hunter-gatherers.

The Americas were also less fortunate than the Old World in the plants that were available for domestication. The most widespread package was a combination of maize, beans, and squash that originated in Mexico. Maize in particular came to be cultivated as far north as Canada and as far south as Argentina, a remarkable success, given that it had to diffuse from one climatic zone to another. Other domesticates, like potatoes in the Andes and manioc in Amazonia, did not spread beyond the types of environment to which they were initially adapted. The relative disadvantage of the Americas became clear when the coming of the Europeans brought Old and New World crops into competition. Several New World crops, like maize and the potato, were widely adopted in the Old World; but the colonization of the New World by Old World domesticates was a much more extensive process.

The relative weakness of New World farming is reflected in the story of its emergence. Currently there are two rival chronologies in the field, a long one and a short one. The long chronology places the beginnings of plant domestication in the early Holocene, with dates going back to 10,000 B.C. The short chronology prefers dates no earlier than the fourth millennium B.C., a good five thousand years later. The choice turns in part on how hard or soft the evidence has to be. Yet for our purposes there is little need to choose between the two chronologies. On the short one, domestication started some five thousand years later in the New World than in the Old; on the long one, it started just as early in both, but brought about no radical change in New World societies for the next five thousand years. Whichever is the case, the upshot was that the Neolithic revolution got under way much later in the

New World. Real villages do not appear in the Americas before the short chronology would lead us to expect them.

Within the New World the cases of the two continents are somewhat different. In North America the only region in which successful farming emerged independently was the Mexican highlands with its classic package of maize, beans, and squash. Maize at least was domesticated in or by the fourth millennium B.C., though village life appears only in the second millennium.

From the Mexican highlands this package, and maize in particular, spread northward over considerable distances to two regions of what is now the United States. One was the southwest, a territory with no earlier domesticates. Here farming arrived in the second millennium B.C., but did not become predominant for another couple of millennia. The other region was the southeast, where maize arrived in the first millennium A.D.; this region had possessed some marginal local domesticates since the third millennium B.C., but it was not until some time after maize had been adopted that farming brought about major changes in society. In due course maize cultivation diffused widely in the eastern half of the United States.

As we have seen, maize also spread to South America; it was cultivated in Ecuador in the second millennium B.C. But by this time farming was already well established in the central Andes. Here the domestication of the main plants and animals had probably happened in the fourth millennium B.C. The major crops in the Andean region were a grain called quinoa and the potato; the major animal was the llama. The impact of farming on society is clearly visible in the third millennium B.C. The walls that appear in figure 2 are in fact located at Caral in the coastal lowlands of Peru; here domesticated crops and monumental architecture date from about 2600 to 2000 B.C.

Amazonia is the joker in the pack, since in tropical forests good archaeological evidence is hard to come by. Domesticated manioc may—or may not—have as long a history as Andean crops. The tropical horticulture of the Amazonian lowlands was most successful when practiced on or near the floodplains of the major rivers of the region, which in turn provided obvious avenues for its diffusion. Indeed, farming peoples with canoes seem to have moved not just down the rivers of northern South America but also through chains of islands offshore. Thus the linguistic affiliations of the Taíno, who occupied the Greater Antilles at the time the Spanish arrived, point to an origin deep in the interior of tropical South America.

American Civilization

If the weakness of American farming was good for the conservation of the hunter-gatherer way of life, it was bad for the emergence and spread of civilization. At the point at which the Old World and the New came into definitive contact in 1492, civilizations occupied a vast territory in the Old World, but only a small part of the New. In fact, just two areas in the Americas supported civilizations, though both had done so for a considerable time. One was Mesoamerica, a region combining the Mexican highlands with the adjoining lowlands. The other, if we overlook the absence of writing, was the central Andean region, including the coastal lowlands—roughly the modern country of Peru plus some territory to the north and south of it. Though the peoples of this region lacked writing, they did have something less satisfactory in lieu of it (as we will see in the final section of this chapter). Outside these two regions there were no writing systems at all, no cities, and no states. The richer farming societies might be ruled by powerful figures whom we can comfortably, if a little vaguely, call chiefs. In the southeast of the United States, for example, some of the large mounds of earth that survive to this day are known from early European testimony to have had chiefly residences on top of them. Moundville in Alabama has some twenty major mounds dating from the thirteenth or fourteenth century A.D.; these earthworks attest the considerable power of chiefs to mobilize the labor of their subjects. Similar societies were to be found in South America, notably in the northern Andes. But none of them were in the same league as the Aztec or Inca states.

A basic fact about the two civilizations of the New World is the absence of relations between them, whether by land or by sea. This did not preclude the diffusion of a couple of important innovations from one region to the other. As we saw, maize (but not writing) reached the Andean region from Mesoamerica; and a measure of metalworking (but not the llama) spread in the opposite direction at a much later date. Yet there were no direct relations between the two civilizations, and they would appear to have been ignorant of each other's existence. One reason for this was the relatively low level of New World maritime technology. On the Pacific we hear of canoes, rafts, reed boats—but nothing to compare with the ships that enabled the Spanish to link the two regions in the course of a mere fourteen years. Another reason for the lack of contact was that the civilizations of the Americas were separated by latitude, not by longitude as in the case of Eurasia. Hence any linkages between the two regions had to cut across the climatic bands.

This means that the context of the New World civilizations was marked not just by the absence of technologies taken for granted in the Old World but also by a much greater degree of isolation. It may be worth reviewing a few features of these civilizations against this background.

We can start with something that in itself needs no special explanation: the absence of a strong tradition of long-term political unity in either of these multi-ethnic civilizations. In Mesoamerica the Spanish found an uneven patchwork when they arrived. The Aztec empire in the highlands was extensive, but there were significant territorial gaps in it; the Mayan lowlands lay outside it, and were divided into a large number of small states. There were precedents for both configurations. In the highlands the Toltecs were remembered to have ruled an empire a few centuries earlier (it may have fallen in the twelfth century), and archaeology suggests that the city we know as Teotihuacán was the center of an empire in the first centuries of our era. In the lowlands the Mayan inscriptions of the first millennium A.D. demonstrate a political fragmentation comparable to that encountered by the Spanish. In short, large-scale political organization seems to have been intermittent, and a feature of the highlands rather than the lowlands. The Andean region looks very different to us because the arrival of the Spanish coincided with the zenith of the Inca empire, a massive and centralized imperial state that had recently conquered the entire region, highlands and lowlands. But we have no reason to think that a state of such dimensions had ever arisen there before, though there had certainly been states of considerable size and power.

A more interesting, because more distinctive, feature of the New World civilizations is that in neither case was there a core ethnic group identified as the originators or continuing proprietors of the civilization as a whole. It is likely that the pioneering role in the development of Mesoamerican civilization was played by a people of the western lowlands whom modern scholars have chosen to call the Olmecs. Olmec culture was already taking shape toward the end of the second millennium B.C., considerably earlier than comparable developments elsewhere in Mesoamerica (though the first evidence of writing comes not from the Olmecs but from the Zapotec area in the highlands, where it dates from the middle of the first millennium B.C.). But Olmec culture lasted only about a millennium, and no memory of it was preserved among the historically known peoples of Mesoamerica. Instead, at the time the Spanish arrived, the civilization consisted of a loose family of cultures, each of which was embedded in its own ethnic context, and no one of which played a central role. There was no such thing as a shared classical language, for example; and we have no reason to think that matters had ever been very different. In the Andean region the highland Chavín culture of the first millennium B.C. might be seen as playing the part of the Olmecs (though monumental architecture goes back to the third millennium). But again, there is no later memory of the culture, and no clue to the ethnicity of those who developed it.

This brings us to a final feature of the New World civilizations that is worth considering here: the relative shallowness of their historical memories.

In the Mesoamerican case the issues are murky, but scholars tend to see the peoples of the region at the time of the Spanish conquest as in possession of historical memories reaching back to perhaps the tenth century A.D. Thus in the highlands the chronicles of the Aztecs preserved an impressive record of their own history going back about a century. They also knew something in historical, and not just legendary, terms about events going back a few centuries before that; and though much of this material was likewise about their own history, they had some conception of the role of the Toltecs as well. Yet they had nothing to tell of the people who created the imperial city of Teotihuacán, let alone their predecessors; our only information there is archaeological. In the lowlands the Maya of the early sixteenth century preserved a record of events comparable to that found in the highlands, but the detailed, if fragmentary, information we possess on Mayan history in the first millennium A.D. is overwhelmingly derived from the monumental inscriptions of the period, and not from chronicles still in circulation when the Spanish arrived.

In the Andean region historical memory was significantly shallower. The Incas put a great deal of effort into remembering their own history, in effect establishing foundations to preserve a record of the life and deeds of each of the Inca rulers, at considerable cost in revenues and personnel. This was not a disinterested activity: Inca history was closely related not just to the prestige of the Inca state as a whole but also to that of the particular Inca lineages associated with each ruler. Not surprisingly, the Incas had no interest in extending this high-maintenance historiography to their predecessors. Thus Tiahuanaco in the southern highlands, a plausible imperial center comparable to Teotihuacán and of roughly the same antiquity, is likewise known to us only by its ruins, and the same goes for Huari farther north. The rule in the Andean region is that the only state history we possess is Inca history. The sole exception to this is the state of Chimú in the coastal lowlands, a highly centralized kingdom conquered by the Incas about 1470. The Incas did nothing to preserve its history directly, but they did allow the dynasty to continue to exercise a measure of power under their overlordship, and as late as the beginning of the seventeenth century the royal family still retained a role under Spanish rule. It is doubtless to this survival of the dynasty that we owe the short, but valuable, account of its history that has reached us through the Spanish sources. In the Andean case, moreover, writing was not available to counteract the fading of historical memory in pre-Columbian times.

As we already indicated, it seems likely that a common thread in the background to these features of the New World civilizations is their isolation. Neither was in contact with other civilizations that could have served as models or rivals, and thereby acted as a stimulus to political unification and a heightened sense of identity. Indeed, in each case the idea of a single overarching civilization is a product not of the native cultures themselves but of our modern understanding of them. This understanding, however, is not obviously wrong. In fact the next section will show that, in the case of one key institution of Mesoamerican cultures, our understanding is unquestionably right.

II. Mesoamerican Calendars

We take our calendar for granted and do not usually think much about it. The result is that we tend to have little sense of its history, or of the way in which it relates to the set of all possible calendars, or even of the importance of having a calendar at all. One way to stop taking our calendar for granted is to see it against the background of a calendar type that once prevailed over much of the Old World. The elements are familiar: days, months, and years. What is unfamiliar to us about this type of calendar is its insistence that the length of the month be keyed to the phases of the moon. The major problem this creates lies in the relationship between the month and the year: twelve lunar months do not contain enough days to make a year, but thirteen contain too many. The solution is to vary the length of the year, so that roughly one year in three has thirteen months, while the rest have twelve. If you do not like this variability, you have to break the relationship between the lunar month and the year. The Muslims are unique in that they made this break by dispensing with real years: they take a block of twelve lunar months and choose to call it a "year." The Europeans, following the ancient Egyptians, take another tack: they divide the year into twelve units that bear no relationship to the phases of the moon, but persist in calling them "months." There is obviously a certain inertia at work here. Once you have severed the link with the moon, there is no good reason why the number of your pseudo-months should continue to be twelve, or their length to approximate that of a real lunar month.

Another system of timekeeping that we take for granted is the week, a cycle of 7 named days; this too was widespread in the Old World long before modern times. What does not occur to us is to take two such cycles and run them concurrently. For example, suppose we were to set up a 2-day cycle and a 3-day cycle. Call the days of the first cycle "1" and "2," those of the second "red," "green," and "amber," and run the two cycles together for 6 days:

Day 1:	1	red
Day 2:	2	green
Day 3:	1	amber
Day 4:	2	red
Day 5:	1	green
Day 6:	2	amber

Here the simple 2- and 3-day cycles generate a complex 6-day cycle in which each of the 6 days has a distinct designation ("1 amber" and the like). This way of doing things may seem strange, but it is well established at the eastern end of the Old World. The Chinese, in particular, combine a 10-day cycle with a 12-day cycle to generate a 60-day cycle (60 being the lowest number divisible by both 10 and 12). This cycle is vital to accurate chronology in Chinese history: the months slither around, but the 60-day cycle has been working like clockwork since time immemorial.

We are now equipped to take on the calendars of Mesoamerica. Since life is governed by the seasons, in the New World just as in the Old, it makes sense to look for a recognizable Mesoamerican concept of the year. Reassuringly, we have no trouble in finding it: a year of 365 days, just like that of the ancient Egyptians. Of course, in astronomical terms this is very slightly too short, a problem Julius Caesar more or less solved for us with his leap-year system; but the Mesoamericans lived with this discrepancy. Assuming that their minds worked like ours, we would then expect them to divide their 365-day year into smaller units; and again they oblige. Just like the ancient Egyptians, they set aside 5 days at the end of the year; this is a good idea, since 360 is a nicely divisible number. What the Egyptians did with their 360 days was divide them into twelve 30-day months—an unimaginative choice, but a distinctly tidier system than the one we have today. The Mesoamericans, by contrast, divided the 360 days into eighteen units of 20 days each. Among the Maya of Yucatán, for example, the 7th of Pop is the 7th day of the 20-day "month" called Pop. The number 20 was very much in place in Mesoamerica because the counting system was vigesimal (in other words, the base of the system was 20, not 10 as it is with us). Though the resulting structure of the 365-day year may strike us as a little bizarre, it is not baffling.

But the Mesoamericans did not leave it at that. Alongside their 365-day year, they had a cycle of 260 days. This cycle was complex: it was generated, like the hypothetical 6-day cycle set out above, by running two simple cycles concurrently. One of these was numerical, and ran from 1 to 13; the other was a cycle of 20 day names. So 7 Coatl, in the calendar of the Aztec capital of Tenochtitlán, was not the 7th day of a "month" called Coatl; Coatl was not a "month" of any kind, but rather the 5th in the cycle of the 20 day names. You get to 7 Coatl by running both cycles until 7 and Coatl come together, which happens on the 85th day of the 260-day cycle.

Despite their having quite different internal structures, the 365-day year and the 260-day cycle can be used conjointly. One way to do this involves the relationship between the 260-day cycle and the beginning of the 365-day year. The key point is that for any given Mesoamerican calendar there are 52 and only 52 days out of the 260 on which the beginning of a year can fall (if you like this kind of thing, you can work out just why this is so). The upshot is that for 52 years, we can distinguish each year by the day in the 260-day cycle on which it starts; once 52 years are up, we are back where we started. So we now have a grand cycle of 52 years (or 73 cycles of 260 days, or 18,980 days). The use of this cycle in historical records lies behind the hard-edged chronology of the accounts of recent Aztec history that were composed after the Spanish conquest; there is no such precision in the corresponding narratives of recent Inca history. For more ancient history, of course, the system does not work so well—Mesoamericanists are forever worrying about which 52-year cycle a given event should be assigned to (did it happen in 1204 or 1256 or 1308 . . . ?).

This analysis has presented Mesoamerican calendars as if they were purely functional. In fact, they trailed large amounts of cultural baggage that brought the calendar into relation with Mesoamerican religion and cosmology. In the Aztec area each day of the 20-day cycle had its supernatural patron, and each day of the 260-day cycle was coded lucky, unlucky, or neutral. A particular day might be both lucky and unlucky for different reasons; it took a professional diviner to resolve such complexities (no doubt for a fee). The 5 leftover days at the end of the year were dangerous; it was a good time to stay at home. There were similar conceptions among the Maya; thus days in the 260-day cycle might be spoken of as if they were intelligent beings. Incidentally, the 260-day cycle survives among several Mayan ethnic groups today, complete with widely diverging ideas about which days are lucky, unlucky, or neutral. All this is comparable in its arbitrariness to the traditional English superstition about Friday the thirteenth, but far more systematic.

The basic shape of the calendar did not vary greatly from one Mesoamerican people to another. In the west the Tarascans at the time of the Spanish conquest lacked the 260-day cycle. In the east the Maya of the first millennium A.D. used a "Long Count," which gave absolute dates in an era starting (for no known reason) in or around 3114 B.C., thus avoiding the long-term ambiguity of the 52-year cycle. But these are unusually large deviations. For the most part the divergences are of a kind one would expect, given the expanse of space and time over which this calendar type prevailed—the surprise is rather that its basic structure should have been so uniform. Given this relative uniformity, and the fact that the Mesoamerican pattern has no precise parallels elsewhere in the world, it is clear that we have to do with a case of diffusion. Like Australian subsections, this is a system that must have originated at some particular place and time, and subsequently spread to other regions. It is in character with Mesoamerican civilization at large that we have no idea when or where to place these origins (though we know from inscriptions that a calendar of the relevant type was already established in the Zapotec area of the Mexican highlands in the middle of the first millennium B.C.). Nor do we know how to imagine the process by which the system spread. What we can say is that in each case the calendar as we know it seems to be pretty much at home in its local ethnic environment. Thus where names of days or "months" are involved, each people has its own. And the calendars of the various ethnic groups were in no way synchronized; even within the Valley of Mexico, those of different towns seem not to have been in phase in Aztec times.

This family of calendars provides a good example of a phenomenon widespread in human cultures. Few societies can do without a calendar of some kind, and a complex society needs a reasonably precise one. Once it possesses such a calendar, it may have to adjust it from time to time, but there is no need to embroider it. Our own calendar is a case in point: it works, and for the most part that is enough for us. But cultures have a way of picking on some aspect or other of their pragmatic arrangements, and elaborating them in respects that have no obvious utilitarian justification. This seems to be the case with Australian subsections; it is undoubtedly so with Mesoamerican calendars. What we see here is again a human propensity for gratuitous cultural embroidery. The reason the example is a good one is simply its dramatic visibility to anyone coming from a Western culture: it so happens that our restraint in calendric matters contrasts sharply with the extravagance of the Mesoamericans.

Yet these same calendars can also be used to illustrate the limits of cultural diversity among humans. A Mesoamerican calendar is immediately recognizable for what it is—a calendar, not some exotic practice bearing only a faint resemblance to what in our culture is called a calendar. Moreover, it is quite obviously a calendar developed by people living on the same planet as ourselves: it takes the day for granted as the basic calendric unit and constructs a year of 365 days. Where we have trouble grasping the workings of these calendars, the reason is merely that they are intricate and unfamiliar; they are far from being so deeply alien to us that we do not know how to begin to understand them.

The fact remains that Mesoamerican calendars stand significantly apart from those of the civilizations of the Old World. The Old World calendars resemble each other more than any of them resemble those of Mesoamerica. This is interesting: it suggests that Old World calendars owed more to each other than we can ever hope to prove from the specific evidence available in particular cases. By the same token, the distinctiveness of Mesoamerican calendars suggests the effects of the emergence of a civilization in isolation from others.

III. The Quipu

Unlike Mesoamerican calendars, the object from the Andean region shown in figure 12 is purely utilitarian. The material of which it is made is string and nothing else, which makes it easy to carry around—several such objects can be stuffed into a cloth bag. At the top is the main string, to which in this case over ninety pendent strings are attached; some of these in turn have substrings tied to them. The result is elaborate, but hardly a work of art. Normally such an object would not survive when no longer cared for by its owner, but this one (like a good many others) owes its excellent state of preservation to the aridity of the coastal lowlands of Peru, an environment in which ancient textiles also survive in some quantity. The object is a quipu and was doubtless taken from a late pre-Columbian burial—even Inca rulers were buried with quipus.

The message of a quipu was in the knots. Beyond this, only two things seem reasonably certain about the way quipus worked: that they recorded numbers, and that they did not record words. The recording of numbers was relatively straightforward. Several distinct knots were used to encode them, and they did so in the same decimal system that we use. This is not speculation. We sometimes find what is called a top cord grouping a set of pendent strings; in such cases, the number on the top-cord checks out as the total of the numbers on the pendent strings, thus confirming our reading. By contrast, neither the quipus we have in our hands nor the accounts left by Spanish observers support the idea that the knots were a form of writing—that what they did was reduce spoken language to a visual code.

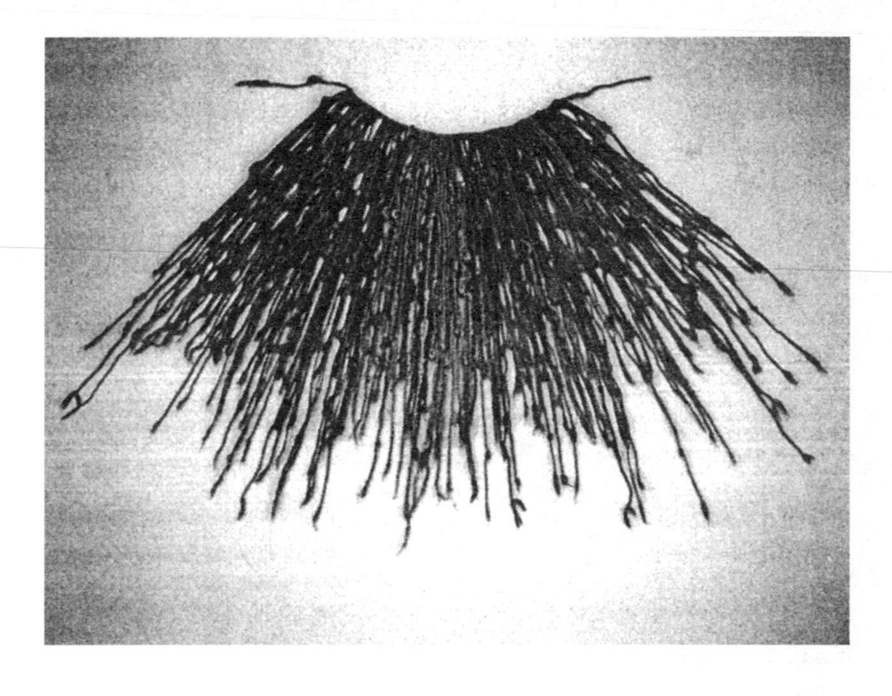

Fig. 12: A quipu.

Unfortunately this leaves us with a large gray area. One might imagine that a quipu contained nothing but numbers, and that its owner had to rely entirely on his memory to know what each number referred to. But this is unlikely. Pendent strings are attached to the main string of a quipu in a certain order; substrings are attached to some pendent strings but not to others; above all, many quipus (including the one shown here) have strings in a variety of colors. All this presumably meant something to experts. But does this mean that such coding enabled any expert to read any quipu even if he had never seen it before? This is hard to imagine, and one Spanish source tell us explicitly that it was not the case. In other words, it seems that understanding the numerical contents of a quipu required memory specific to that quipu.

This does not preclude the idea that quipus could have been used as devices to prompt memory of nonnumerical material—a highly elaborate version of tying a knot in one's handkerchief. In this connection many Spanish sources mention the use of quipus in connection with the recording of historical narratives, and they may well be right. What we cannot tell, of course, is exactly how much of the information in an oral rendering of the narrative could actually have been encoded on the quipu. Still less do we have any understanding of the code itself.

One thing that is abundantly clear is that quipus were essential to the functioning of the Inca state. They were used to record the many kinds of statistical information needed for administrative purposes: census data, labor services, quantities of goods in storage along the roads, and the like. One early Spanish observer asked a chief to explain the system to him; the chief showed him quipus recording all the goods he had delivered to the Spanish since their arrival.

The Incas do not seem to have invented the idea of the quipu, and it certainly outlived them. Decades after the conquest, quipus remained in use for administrative purposes within the native society. In fact, the quipu was still alive in remote areas in the last century—for example, on the islands of Lake Titicaca. But once the Spanish had introduced writing to Andean society, the quipu was bound to lose ground.

What is interesting about the quipu is precisely this inverse relationship to writing. According to Chinese tradition, knotted cords were used for administrative purposes in early China, but the sages later replaced this practice with writing. Society in the Andean region—or, more particularly, the Inca state—had likewise reached a level of complexity at which it needed to collect and conserve large amounts of information, far more than people could comfortably carry in their heads. Unquestionably the most effective solution would have been writing, the technology adopted not just in China but in all Old World societies at a comparable level of complexity (not to mention those of Mesoamerica). But Andean civilization had no access to an already existing writing system, and it did not invent one for itself. Instead, it made do with a workable system for recording numbers, but one that in other respects must have placed a significant burden on the memories of those who operated it. In short, the Incas lacked something they needed.

This lack was not confined to the immediate needs of the bureaucracy. It was also apparent in another domain we have touched on, the recording of historical information. Around 1600 an author writing in Quechua, the language of the Inca empire, lamented the previous absence of writing among his people in these terms: "If the ancestors of the people called Indians had known writing in earlier times, then the lives they lived would not have faded from view until now. As the mighty past of the Vira Cochas [the Spanish] is visible until now, so too would theirs be." This author naturally made his point by drawing a contrast with the Spanish; but he could also have referred to Mesoamerica. It is thanks to the fact that the Maya had writing that they are the only American people whose history in the first millennium A.D. did not permanently fade from view.

In sum, the quipu is a lot less stimulating to the cross-cultural imagination than Mesoamerican calendars, but it has a significant, if by now familiar, implication. It demonstrates the effects of isolation. Isolation can make for interesting diversity, as in the case of Mesoamerican calendars: they are refreshingly different from Old World calendars, and they are not on balance any the worse for it. But isolation also insulates a culture from a wider field of competition and stimulus; and in the long run this is unlikely to be to its advantage. Thus all the early Old World civilizations possessed writing, whereas only one of the two American civilizations had it. What this surely reflects is the interconnectedness of the Old World, even in very ancient times, and the isolating geography of the New World throughout its independent history. The fact that the one civilization that ever developed in the Southern Hemisphere neither invented nor acquired writing shows how hard it was to make history in pre-Columbian America. No wonder Atahuallpa, the last Inca emperor, was fascinated by the ability of the Spanish to read and write. It is said that, during his captivity before his execution, he devised an experiment to establish whether this ability was innate or acquired; he went to his death knowing that the skill was culturally transmitted.

CHAPTER 6

AFRICA

I. The African Cultural Gradient

Africa is the one continent where we do not have to wrestle with the date of the arrival of modern humans: they evolved there. South of the Sahara this antiquity carries with it a much greater genetic diversity than is found among the human populations of the rest of the world. This diversity includes the most drastic, or at least the most conspicuous, physical adaptations of any human populations to climatic conditions: the short stature of the Pygmies in the hot and humid rainforest, and the tall, thin body build of a scattering of peoples of the hot and dry savannas. Yet in Africa, as elsewhere, the prime factors explaining the diverse trajectories of human societies are more directly environmental. What sort of an environment does Africa provide for humans?

It makes some sense to think of Africa as a southern continent. Its current position on the globe is to the south of western Eurasia, and in origin it is the largest fragment of the old southern supercontinent of Gondwana. Yet Africa is not as southern as we tend to think. It actually extends slightly farther to the north of the equator than it does to the south, and it is about twice as wide in the north.

This location gives Africa a climatic symmetry comparable to that of the Americas, but considerably more limited in scope (see map 4). As in the Americas, there is a substantial tropical belt around the equator characterized by rainforest, though the amount of it is far less generous. To the north and south of this band there is open country, which may be grassland where the summer rainfall is adequate or desert where it is not. On the northern side the grassland forms a belt immediately to the north of the rainforest; still farther north lies the Sahara, the world's largest and hottest desert, stretching continuously across the continent at its widest (though this region was significantly more hospitable several thousand years ago, when the Holocene climate was warmer than it is now, and consequently wetter). On the southern side of the rainforest the desert areas are in the south and west. Beyond the deserts lie the two extremities of the continent that enjoy a Mediterranean climate, with its winter rainfall. To the north Africa ends in a long coastline; while about half of this is desert, the other half has enough rainfall to share the climate of the Mediterranean at large. To the south a small region at the bottom of Africa has a climate of the same Mediterranean type.

As with the Americas, this picture of climatic bands is confused by the presence of mountains. But there the resemblance ceases. Africa is a relatively undisturbed piece of continental crust, remote from subduction zones except on the north, so it is generally flat. Such mountains as it has are concentrated in two regions. One is the northwest, where the Atlas Mountains are of the same recent vintage as the Alps, and result from Africa's collision course with Europe. The other, much larger region is East Africa, where the formation of mountains is associated with the splitting of the earth's crust that has given rise to the rift valley. The rifting runs from north to south, and is one branch of a massive system of

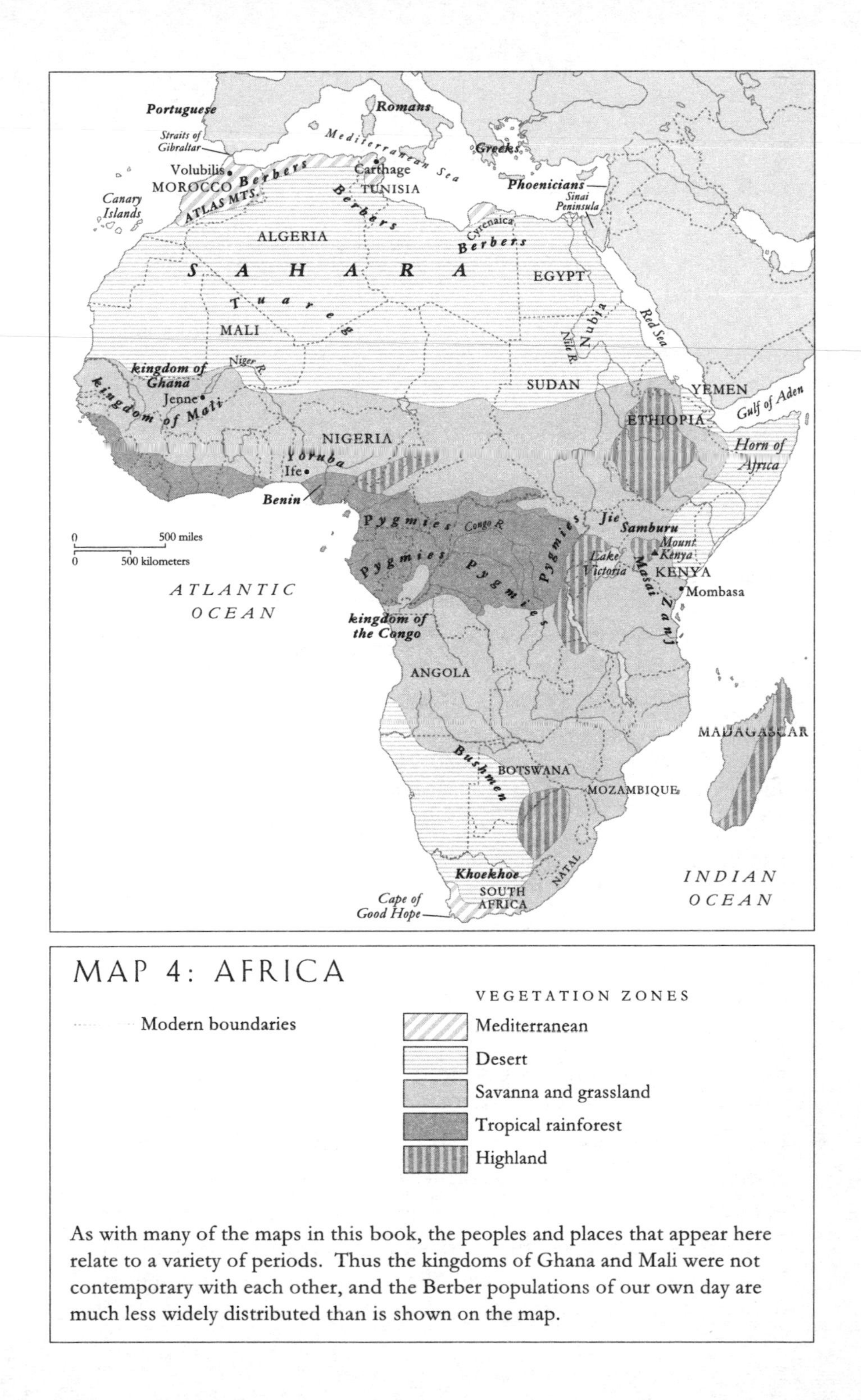

MAP 4: AFRICA

As with many of the maps in this book, the peoples and places that appear here relate to a variety of periods. Thus the kingdoms of Ghana and Mali were not contemporary with each other, and the Berber populations of our own day are much less widely distributed than is shown on the map.

faults; the two most conspicuous branches of the system are the Gulf of Aden and the Red Sea. So whereas the Americas have their mountains in the west, Africa's are mostly in the east; and their size does not bear comparison with the massive ranges of the New World.

A final major feature of African geography runs from south to north: the Nile. Unlike such rivers as the Amazon or the Mississippi, the Nile brings water to a part of the world that desperately needs it. The Mediterranean is markedly stingy with its rainfall along the eastern half of the northern coast of Africa; Egypt, if left to the mercy of the Mediterranean, would be an unrelieved desert. What the Nile does is to give Egypt a transfusion of East African rainfall that has its origin in evaporation off the Indian Ocean.

Several features of this layout are crucial to the place of Africa in history. One is the proximity of the northern part of the continent to the western half of Eurasia. Africa is joined to it at the Sinai Peninsula, and comes close to it at the Straits of Gibraltar and the mouth of the Red Sea. Moreover, what separates northern Africa from western Eurasia is mostly the Mediterranean, which has been a zone of high maritime interaction over the last few thousand years. This means that the northern end of Africa is far from isolated, in sharp contrast to Australia or the Americas. Not so the southern end. This too shares a climate with other regions in its hemisphere—central Chile and western Australia—but it is utterly remote from them. Until the development of navigation on the oceans, any contacts between the southern fringe of Africa and a wider world had to cross the climatic bands of the continent. Here the open grasslands are good for interaction, but the deserts and the dense rainforest are not. We can add to this that the mountains of East Africa are not a corridor comparable to the Andes. In short, we can expect the history of the continent to be marked by a steep cultural gradient, with the advantage going to the north. And this is indeed what we find.

Hunter-Gatherers and Farmers in Africa

In southern Africa numerous small hunter-gatherer populations still survive, or did so until recently. In the open country are the Bushmen, brown rather than black populations such as the G/wi of Botswana—the "/" in their name is a dental click, such as English-speakers use to express annoyance. These groups are widely scattered, but much more frequent in the desert areas of the center and west than in the grasslands of the east. Their languages share the frequent use of clicks like "/"—a phenomenon unknown outside southern Africa. This is reminiscent of the common features of the languages of Australia; and in the same way, the languages of the Bushmen cannot convincingly be reduced to a single family. This suggests that Bushmen have been in the region for a long time, and from any historical perspective we can think of them as aborigines. To the north of the Bushmen are the Pygmies, the equally aboriginal hunter-gatherers of the tropical forest. Some twenty different groups are known; presumably they once had languages of their own, but they now speak those of farming populations with whom they have or once had relations. To the east are numerous small hunter-gatherer populations in the East African highlands. Taken as a whole, these scattered hunter-gatherer populations of southern Africa look like residues of a time when the region was as much a preserve of hunter-gatherers as Australia before the eighteenth century. As we will see, that time, though not so recent as in the case of Australia, was not more than two or three thousand years ago. Against this southern pattern we can set the virtual absence of hunter-gatherers from northern Africa, even in ecologically similar regions. They hardly existed in modern times, and they have not been conspicuous for some thousands of years.

Unsurprisingly, the origins of African farming lie in the northern half of the continent. One development that may have taken place very early is the appearance of domesticated cattle in the Sahara, then a more benign environment than it is today; in the eastern Sahara this may go back to around 7000 B.C. As we saw in chapter 2, this could have been an independent domestication. On the Mediterranean coast to the north of the Sahara, it was the Near Eastern farming package that spread, though at a somewhat later date. In the savanna belt to the south of the Sahara, domesticated animals were introduced from the north, but in a region of summer rainfall the plants had to be local domesticates; here the establishment of farming seems to have taken place within the period 4000–1000 B.C. Farther south, in the tropical forest, a pattern of agriculture appeared that was centered on the cultivation of domesticated yams; but as in Amazonia, the chronology remains murky. Thus the appearance of farming in Africa shows some clear instances of diffusion from the Near East, but often it is hard to decide between stimulus diffusion and independent development. By contrast, an aspect of material culture in which the African savanna had clear priority over the Near East was the development of pottery around 9000 B.C. (as against about 7000 B.C. in the Near East).

The story of the spread of farming to southern Africa is a much more recent one. Here there is no sign of independent domestications, with or without external stimulus. In one way this is surprising: the savannas of eastern and southern Africa are the only part of the world in which the wild herbivores of the late Pleistocene still survive on a significant scale. But for whatever reason, none of them were suitable for domestication. Instead, domesticates entered the region from the north. Two distinct processes were involved: the colonization of the south by farmers from the north, and the adoption of northern farming by hunter-gatherer populations already in the south.

The first process is the better known. The evidence for it comes from two sources, archaeology and linguistics. The archaeological record shows that by the second century A.D. there was a farming population between Lake Victoria and the East African coast that also practiced iron metallurgy. This population then spread rapidly down the east coast of Africa, reaching Natal in less than two centuries. It stopped there, presumably because its crops were adapted to the summer rainfall of the tropics, as opposed to the winter rainfall of the south coast. Gradually this or similar populations moved into the interior of southern Africa, including eventually the west; but this last part of the process is very little known.

The linguistic evidence consists in two points. First, virtually all the languages spoken by the black populations of southern Africa are closely related, belonging to a single subfamily known as the Bantu languages. This is quite unlike the situation farther north in sub-Saharan Africa: there the language map of black Africa shows no such homogeneity, and extensive groupings of languages are much harder to establish with confidence. Second, there is good reason to place the homeland of the Bantu languages in this northern region, more precisely in the northwest near the Atlantic coast. It is here that Bantu itself is most deeply differentiated and that the non-Bantu languages most closely related to it are found. Taken together, these points strongly suggest that the Bantu languages entered southern Africa from the north in the relatively recent past, perhaps within the last couple of thousand years.

So the arrival of the iron-using farmers and the spread of the Bantu languages look like two sides of the same coin. But this conclusion, though it may be irresistible, does raise a problem. The archaeological evidence agrees with the linguistic evidence in suggesting a movement from north to south; but whereas archaeology points to an origin in the northeast, linguistics points to the northwest. How do we bring them together? We can reduce the gap a little by noting that an iron-using culture appears to the west of Lake Victoria earlier than it does to the east. But to bridge the rest of the gap, we are reduced to positing an early eastward migration of Bantu speakers in the direction of Lake Victoria, a movement for which evidence is otherwise lacking. Be this as it may, we are still on firmer ground than with Pama-Nyungan.

The other aspect of the spread of farming to southern Africa concerns its adoption by hunter-gatherers already living in the region. Pastoralism (but not cultivation) was at some point adopted by people of the same physical type as the Bushmen. Such brown pastoral populations are known as the Khoekhoe peoples; they are still a significant presence in the central and western regions of southern Africa, and they were more extensive in the past. They speak click languages of a family to which a number of Bushman languages also belong. Archaeologically they are not well known, but their way of life seems to go back a couple of thousand years. The origins of their pastoralism must lie in northern Africa, whence their livestock derives. The simplest hypothesis would be that the ancestors of these pastoralists obtained their livestock from the iron-using black farmers. But there are indications that their pastoralism may have reached the south by a different route.

Literate Culture in Africa

If one major southward movement was the spread of farming in southern Africa, another was the spread of literate culture in northern Africa.

Africa was the home of what was probably the world's second-oldest civilization, that of ancient Egypt. For most of its prehistory, Egypt had been rather a backwater. Farming is not attested in the Nile Valley until the sixth millennium B.C., and it is only in the fourth millennium that Egypt becomes a place to watch. But the development that then took place was rapid, with archaeological evidence of increased social stratification culminating in the emergence of the Egyptian state toward the end of the millennium. From that point on there was a tradition of Egyptian monarchy that was still in place in the fourth century B.C. and survived in a residual form into Roman times.

This monarchic institution played a remarkably salient role in Egyptian civilization, or at least in what we know of it. Like the Narmer Palette (see figure 10), the remains of ancient Egypt give us a great deal of information (or disinformation) about the doings of kings, but tell us much less about anything else. The major exception is the Egyptian way of death, as we will see in the final section of this chapter; but even there, much of what we learn, particularly in the early period, is about the death of kings. At the same time, the roll call of the Egyptian kings lay at the core of Egyptian historical memory. Manetho, an Egyptian priest who wrote an account of the history of his country for the Greeks in the early third century B.C., still had at his disposal an authentic historical record going back to around 3000 B.C.—a time depth that could have been matched only in Mesopotamia. But the tradition of monarchic rule was not, in practice, unbroken. It was interrupted every few centuries by periods of disunity, and more seriously it was increasingly subject to episodes of foreign rule. No such episodes had occurred in the third millennium, and only one in the second; but in the first millennium they became so numerous as to make foreign rule the normal condition of the country. This in turn sapped the foundations of the high culture that had been associated with the Egyptian state. Already in retreat in the time of Manetho, within a few centuries it was dead.

One might have expected a civilization that lasted over two and a half millennia to have found many imitators beyond its borders. Yet this was not the case. As we will see in later chapters, civilizations vary enormously in the degree to which they export themselves, or are imported by others; that of ancient Egypt was more or less confined to the home market. For the peoples of the Near East it seems to have been in some way less eligible than its Mesopotamian competitor. For the peoples of Africa it was not very accessible. The Nile Valley is flanked by desert on two sides, and desert dwellers do not have much use for civilization. Farther up the river to the south were the Nubians, who unlike the desert dwellers did have a need for a high culture; and in contrast to the peoples of the Near East, they were a captive audience for the Egyptians—sometimes literally so. So the Nubian adoption of the culture of their northern neighbors, with or without ethnic customization, has a long history. But Nubia was a dead end; it was not on the way to other territories where Egyptian civilization might have been in demand, such as the grasslands to the south of the Sahara. It may be possible to find traces of Egyptian civilization here and there in sub-Saharan Africa, but there was no instance of the transfer of the culture as a whole.

It was in the first millennium B.C. that literate cultures began to appear in parts of Africa other than the Nile Valley. The origins of these cultures lay outside Africa altogether, and the process that brought them to the continent was colonization. It began when the Phoenicians, maritime traders of the Syrian coast, founded the city of Carthage in what is now Tunisia. Carthage preserved a Phoenician culture (complete with its alphabetic script) from its foundation in the ninth century until its fall in the second century B.C. Toward the end of this period kingdoms were emerging among the native population of the hinterland, and one of these adopted Carthaginian culture. As a result of this interaction a version of the Carthaginian script came to be widely used for inscriptions in the native languages of ancient North Africa; this script was preserved down to modern times by the Tuareg nomads of the Sahara, among whom it was traditionally taught by the women. But there was no lasting native adoption of Phoenician literary culture.

Another example of the process takes us to the Ethiopian highlands in the middle of the first millennium B.C. In this case the colonists had crossed the sea from Yemen, bringing with them their language and culture, which likewise included an alphabetic script. This time the colonial culture was not imitated by independent African societies, and thus did not spread beyond Ethiopia; but thanks to its association with a resilient monarchic tradition, it has survived down to the present day in a distinctly Africanized form. The core of the Ethiopian population still speaks languages derived from that of the Yemenite colonists and uses a version of the script they brought with them from Yemen, where it died out more than a thousand years ago.

Meanwhile, along the Mediterranean coast, the Phoenicians had been followed by the Greeks and Romans. The Greeks began to colonize Cyrenaica, to the west of Egypt, in the seventh century B.C., and thanks to Alexander the Great they occupied Egypt itself in the fourth. The Romans occupied North Africa and took over Egypt in the course of building their empire in the last centuries B.C. The ruins and Latin inscriptions of Volubilis in what is now Morocco date from the early centuries A.D. and show members of a native tribe living a literate urban life in Roman style.

Yet of all the outsiders of premodern times, none had so far-reaching a cultural impact as the Arabs. The rise of Islam in the seventh century A.D. led to the Arab conquest of the northern coast of Africa, from Egypt to Morocco; ultimately it is due to this expansion that Arabic literary culture prevails throughout this area today. But the Arabs also had something that their predecessors had lacked: they were a desert people for whom the Sahara was territory of a familiar kind. They did not in general make it their business to send armies across the desert, but they did establish a degree of trans-Saharan contact that cannot have been witnessed for several thousand years. When we peer through the eyes of the Greek and Latin authors of antiquity, we catch only the most fleeting glimpses of the world beyond the desert; but with the appearance of the Arabs, for the first time in recorded history the Sahara becomes transparent.

An immediate effect of this is to reveal the existence among the black populations of the savanna of a kingdom called Ghana. It was already there in the eighth century; we have no way to know when it was founded, or what predecessors it may have had, though there is archaeological evidence of urban life in West Africa by A.D. 300. In an eleventh-century source Ghana still appears as very much a pagan kingdom: we hear of idols and sorcerers, and of royal burials in which the dead king was supplied with grave goods and accompanied by the men who used to serve his meals. A reminder that we are in Africa, where matrilineal kinship systems are common, is the rule of succession: the kingship went not to the dead ruler's son but rather to his sister's son. There is no sign of any tradition of literacy in the native language. But like many rulers of societies lacking literate culture, those of Ghana clearly saw a use for it. The king's treasurer, and most of his ministers, were Muslims, and presumably literate in Arabic. There was also a Muslim town a few miles from the pagan capital; its existence doubtless reflected the role of Muslims from the north in the trans-Saharan trade.

In the long run the pagan Africans of this region themselves converted to Islam. Thus Mali, a major West African kingdom in the thirteenth and fourteenth centuries, was already a Muslim state. Its people were not, perhaps, very good Muslims. A Moroccan traveler visiting West Africa in the mid-fourteenth century, who comments on the degree of respect shown to women, was shocked when a local Muslim judge casually introduced his girlfriend; she laughed at the traveler's embarrassment. A century and a half later a ruler who took Islam to heart was perturbed by the fact that all the prettiest girls in the city of Jenne walked about naked, even the daughters of Muslim judges. But the people of the West African savanna were now part of the Muslim world, and under pressure to assimilate its mores.

While Islam came to West Africa by land, it reached East Africa by sea. Here too the Muslim Arabs went farther than their predecessors—their accounts of the region reach far into the Southern Hemisphere. As in West Africa an early result of this is to give us vivid images of societies that we could otherwise hope to know only through the veil of archaeology. A tenth-century source describes an East African kingdom among the Zanj in what is now Mozambique; we hear that an unjust ruler would be killed and his descendants barred from ruling. The king had a cavalry force of three thousand men mounted on cattle—an entirely credible report, since this practice is known elsewhere in southern Africa at a later date. In this region south of the Horn of Africa, the interaction between native pagans and foreign Muslims did not result in a general conversion to Islam, but it did lead to the emergence of a Muslim African population along the coast.

What's in the Middle?

This coverage of the African past has done something for the northern and southern parts of the continent, and something for the east coast. But what about the middle? Here we have neither the dramatic confluence of archaeology and linguistics that we found in the south nor the literary sources that illuminate the north and east. Yet two phenomena are perhaps worth highlighting.

One is the pastoralism of the East African highlands. Pastoralists appear on the plains of northern Kenya as early as the middle of the third millennium B.C. In the next section we will take up an aspect of their social organization in recent centuries; though by no means typical of Africa as a whole, it has some interesting things to tell us. At this point what is worth noting is the contrast between the East African highlands south of Ethiopia and the New World at the same latitude. Unlike the Andes, these highlands did not support intensive cultivation, relatively advanced metalworking, urbanization, or state formation, and were not integrated into the long-distance trade of the coast. The main reason for this is no doubt the aridity of the region.

The other phenomenon we should glance at is the evolution of urban life on the margins of the tropical forest in the region of the lower Niger—at first sight an unpromising environment, particularly in the absence of a maritime commercial scene comparable to that of the east coast. Yet the region is famous for some very fine bronze work dating to the centuries before the arrival of the Portuguese. At this time there was already a substantial kingdom in the region, that of Benin; and if we can judge by oral traditions and later political patterns, there were also smaller dynastic city-states, of which one of the few known to us archaeologically is Ife. Just why this area should have supported more complex societies than were found in comparable regions to the east or west is hard to say; it is tempting to compare these West African city-states with those of another people who lived on the margins of a tropical rainforest, the lowland Maya.

Overall, the most distinctive feature of the African scene remains its pronounced north-to-south cultural gradient. To take just one example: in the south, the G/wi in the last century still made do with a number system comparable in its simplicity to that of the Aranda; in the north, by contrast, the Egyptians were already compiling handbooks on methods of mathematical calculation in the early second millennium B.C. Though each culture could be said to have had what it needed for its purposes, the difference in purposes is telling. Africa contrasts with the Americas in showing not just the costs of isolation but also the benefits of being connected.

II. The Age-group Systems of East Africa

We take it for granted that people of different ages behave differently and that this is appropriate. For example, we feel that a man in his thirties should act his age and not behave like an adolescent or an old man. Equally we expect that, as they go through life, people of the same age will in some ways understand each other better than people of different ages.

All this is part of the fabric of our social life, but it is not something that we embody in formal institutions governed by hard-and-fast rules. If we wanted to describe our society, we could think of childhood, adolescence, maturity, and old age as "age-grades" that anyone blessed with a normal life span has to pass through. But if we so desired, we could easily come up with a somewhat different series, particularly in the middle; and within reason we could place the boundaries between successive age-grades where we wanted. Our society simply does not formalize such things. In the same way, we could describe the set of people currently in a given age-grade (for example, adolescents) as a group—an "age-group"—on the basis that such people feel they have something in common. But it would be a pretty insubstantial sort of group in comparison to a football team.

There are some exceptions to this. At the university at which I teach, and in many others, students pass through a series of four age-grades, starting as freshmen and ending as seniors, before graduating into the terminal age-grade, that of alumni. Different formal (and informal) rules apply to each age-grade. At the same time, the set of students currently in a given age-grade is an age-group—the "class of '07," or whatever—which passes successively through the age-grades. In this setting there is little ambiguity about who is assigned to what age-group, who is currently passing through which age-grade, and, in certain respects, what behavior this calls for. But our society as a whole does not work this way, nor does any other within the broad spectrum of modern societies.

In the premodern world, and residually in the modern world, a considerable number of societies operated age-group systems of the kind that we lack. Examples of these systems were to be found in most major regions of the world. But in terms of their number and their importance in social organization, they were particularly prominent in two regions: Africa and Taiwan (where they were characteristic of the aboriginal, not the Chinese, population). In Africa they were common in several parts of the continent, one of them East Africa. Here the tribes with age-group systems were largely but not exclusively pastoral, or from a pastoral background. As an example we can take the Samburu, a pastoral tribe living to the north of Mount Kenya, as its traditional way of life was described in the late 1950s. These tribesmen were divided into clans; they had no real chiefs, but each clan had its elders. In language and culture they were closely related to the notorious Masai, though they lacked their reputation for military ferocity.

In the Samburu system there are three principal age-grades: boyhood, moranhood, and elderhood. As the first term already suggests, females are not part of the system. Turning to the second term, a moran is a young, unmarried man in his physical prime; before government came to East Africa, he and his likes would have been the warriors of the tribe. Entry to moranhood is clearly marked by a cluster of events. First, a boy is circumcised, typically at the age of fifteen or so. This is a very stressful occasion, not so much because of the actual physical pain, but rather because any sign of flinching will bring disgrace on the boy, his family, and his clan. Soon afterward a ceremony takes place within the clan known as the Ilmugit of the Arrows, and at this time the boy swears to his mother that he will not touch food seen by any married woman—one reason why he will now spend much of his time out in the bush. He is now a junior moran, and can apply red ocher to his head and body. A few years later, when he is about twenty, a second Ilmugit takes place; this is the Ilmugit of the Name, and marks the transition to senior moranhood. At this time the age-group to which our moran belongs chooses a ritual leader from among its members, and the age-group itself is given a name. Several years later, when our moran is about twenty-six, a third Ilmugit takes place, the Ilmugit of the Bull. No moran should marry until he has killed his ox for this ceremony (though if he is socially successful, he will previously have had a girlfriend). Typically a man will be about thirty at his first marriage; he is now settling down and becoming an elder. When he is thirty-four or so, the elders of the clan will bless him and his wife; at this point he ceases to observe the food tabus of a moran and has unquestionably become an elder himself. At some time in these years, when the majority of his age-group have married, a final Ilmugit is held, the Ilmugit of the Milk and Leaves. Just as there are subgrades for the moran (junior and senior), so also there are subgrades for the elders (junior elders, firestick elders, senior elders), but for the most part we can leave them aside.

We have followed the career of an individual male through the early age-grades, but it is evident from this account that in becoming a moran he joins a continuing age-group within his clan. The age-mates go through the Ilmugits together, not as individuals—though it is as individuals that they finally leave moranhood. Membership of the group has a strong effect on social relations: age-mates interact with each other as equals, whereas they are expected to show respect for members of older age-groups. One senior age-group in particular exercises a kind of authority over a junior age-group passing through moranhood, namely the group two subgrades above it. The members of this senior group are the "firestick elders" of the junior group, moral guardians equipped with the sanction of a highly potent curse.

This is roughly how the system is described among the Samburu. Doubtless no two systems were exactly alike, even in East Africa; they could be set up in different ways and used for different purposes. Equally there was probably no people with such a system that could not have done without it. There is no known environment in which humans cannot live without an age-group system. This does not, of course, mean that these systems served no purpose. It certainly makes sense to see age-groups as having important social functions, such as channeling the aggressiveness of young men and mobilizing a military force for the tribe. But pastoral (and other) societies elsewhere in the world do not seem to have found it difficult to achieve such purposes in other ways. The medieval Mongols, for example, had no age-groups, but in military terms they were hardly outshone by the Masai. So we have here yet another example of human cultural diversity, and of the way in which societies tend to seize unpredictably on some particular feature of human life and elaborate it in a manner that is in some sense gratuitous. We also see once more the power of neighborhood in setting such cultural trends: whereas in many parts of the world an age-group system would have been an unusual thing to have, in pastoral East Africa it would have been unusual to lack one.

But something more calls for attention here, and it arises from a rule found among the Samburu that relates the son's age-group membership to his father's. This takes us back to the formidable curse that sanctions the moral authority of the firestick elders. Since it stands to reason that these elders would be reluctant to apply such a curse to their own sons, it makes sense to try to exclude these sons from the junior age-group in question. To this end the Samburu insist that a youth must be enrolled in an age-group junior to that to which his father is a firestick elder, even if this often means that he must wait till well over the age of twenty to become a moran.

Seeking such a relationship was typically, indeed peculiarly, East African. Rules relating the age-group membership of father and son are virtually unknown to age-group systems elsewhere. The role of neighborhood is particularly salient here: as with Australian subsections and Mesoamerican calendars, we clearly have to do with something that must have had a unique beginning and then spread, despite the fact that the region affected had neither ethnic nor political unity.

This spread is remarkable, because in the long run an age-group system that includes such a rule is in principle unworkable—or can be made to work only with messy adjustments. Part of the problem has already surfaced in our account of the Samburu: some youths had to wait several years to enter moranhood, a situation that sets culture at odds with biology. The source of the problem is the simple fact that men father sons at a variety of ages. Take two men of the same generation. The first has a son at the age of twenty, and this son in turn has one at the same age. The second man has a son at the age of forty. The two sons are now effectively of different generations, while the grandson of the first man and the son of the second are age-mates. Yet if we want to maintain a straightforward relationship between the age-groups of a man and his son, it is the two sons whom we want to place in the same group. So when do we form the group? If we delay its formation until the son of the second man is ready to join, the son of the first man will be well past his prime. But if we arrange things to suit the son of the first man, then the fun will be over before the son of the second man is ready to join. If we prefer to split the difference, we will have both problems on a smaller scale: one son will be a bit too old, the other a bit too young. And whichever we choose, the problems are set to get worse with the passing of the generations. In one Ethiopian tribe we hear of initiation ceremonies at which infants and eighty-year-olds joined the same age-group.

The ethnographic record shows that these problems were endemic in East Africa, though for some reason it was in southern Ethiopia that they led to the most serious disarray. We also have evidence that the problems bothered people and that tribes resorted to a variety of devices to resolve or contain them. They might seek to delay the age at which men married, or to prevent them from fathering children after a certain age; some tribes even resorted to infanticide to implement such rules. Or they might loosen the rules a bit, as the Samburu did by allowing a son to enter *any* age-group junior to that to which his father is a firestick elder; this solved the problem of the son born late, though not that of the son born early, who still had to wait.

So far as we know, none of these East African societies had professionals whose role it was to take care of such problems. One tribe whose system was in trouble planned to send to another for advice; but in the end most, if not all, the tribes must have been in the same boat. Like the Aranda, they had elders. Elders can be experienced and wise, but they are not professionals. In this respect the difference between the Samburu and the Aranda was simply that the Aranda were not trying to operate a system with an inherent flaw. The Mesoamericans, by contrast, do seem to have had experts who understood the mathematical and astronomical foundations of their calendar. Presumably it was the business of these experts to advise in the event that the calendar went awry—just as expert opinion was crucial to the making of the executive decisions that gave us the Julian and Gregorian calendars. But in the East African case it is hard to see what even professional experts could have done, beyond a more judicious application of the various corrective devices that were already found among the tribes.

All in all, East African age-group systems reveal more than just the diversity of human cultural choices. They also provide a striking demonstration of the tenacity with which a society can adhere to an unworkable project that it has devised for itself.

III. *Shabtis* and the Egyptian Way of Death

If there is an afterlife, it may be more or less similar to the life we have now, or it may be utterly different. If the second is true, there is little we can do to imagine it, let alone to anticipate what we will need in it. But if the afterlife is reasonably close in character to our present life, we can plan ahead to maximize our future comfort in it, just as prudent people do for old age. The Tarascan king who took with him his chambermaid, his cook, his feather dresser, and so forth clearly believed in such an afterlife, or at least thought it worth betting on. So did Egyptian kings; we have already encountered the slaughter that accompanied some early royal deaths in Egypt. In fact, anyone who is rich and powerful in this world seems likely to need servants in the next world; civilization means inordinate amounts of hard work, and the afterlife would be grim if one could not continue to delegate it. Meketreᶜ, an Egyptian noble who lived around 2000 B.C., was buried with a fine array of

Fig. 13: A typical *shabti* with the spell written on it.

model servants made of wood. His real servants had the good fortune to survive him.

It would therefore be natural to assume the Egyptian statuette shown in figure 13 to be of the same type. Such objects are known as *shabti*s, a word of obscure origin. They begin to be found in tombs of the first centuries of the second millennium B.C., but they differ from the models just mentioned in a couple of respects. They are dressed as mummies, not as living people, and they may be inscribed with the title and name of the person in whose tomb they are placed. It appears, then, that their role is to stand in for the deceased—but in what capacity? Fortunately for us, it seems that *shabti*s, like servants in general, could not entirely be trusted and that magical means had to be used to ensure that they did their job. Starting in the eighteenth century B.C. we find *shabti*s on which a spell is inscribed, and in the following centuries this became common practice. What the spell highlights is a prospect even more disquieting than having to toil to make a living for oneself: the deceased might be called upon to perform labor services in the domain of the god of the afterlife, be it cultivating the fields, irrigating the shores, or transporting "sand of the west or of the east." Should this happen, the *shabti* is to report for duty in place of the deceased, saying, "I will do (them); here I am." From the fifteenth century onward, *shabti*s were further equipped with the necessary agricultural tools. This *shabti* spell is also familiar to Egyptologists from elaborate collections of useful spells for the afterlife that came to be written on papyrus rolls and placed in tombs with the dead. It is the contents of such rolls that we (but not the ancient Egyptians) customarily refer to as the Book of the Dead.

*Shabti*s remained a key element in a decent Egyptian burial for many centuries. Their numbers multiplied enormously, with the result that today they are among the most common of ancient Egyptian antiquities. They came to be mass-produced, with adverse effects on their artistic quality. Instead of being buried with just a single *shabti*, the deceased would be provided with whole teams of them. One document dating from the early first millennium B.C. shows a son purchasing 401 *shabti*s from a temple for his father's tomb: 365 ordinary *shabti*s (making one for each day of the year) and thirty-six foremen. But in the period after the end of native Egyptian rule in the fourth century B.C., *shabti*s gradually disappeared.

Every human society has to have a way of death, and even the Samburu were no exception. But they did not make a big deal of it. Even when an influential elder died, there was no elaborate ceremony; the disposal of the corpse was simple and discreet. So it is unlikely that Samburu practices will leave much for future archaeologists. Fortunately for us, the ancient Egyptians took an extravagantly different view of things; indeed, their remains suggest to us that they spent their lives obsessed with death. There is, of course, an element of illusion here. The Egyptians led their lives in the agricultural lands of the Nile Valley, where ancient remains survive badly, but moved in death to the adjoining deserts, an archaeological paradise. Yet without any question, they took their afterlife very seriously.

The contrast we see here between the Egyptians and the Samburu relates not just to the funerary practices of the two peoples but also to their politics. Whereas the Samburu still did without chiefs in the twentieth century, the Egyptians already had kings five thousand years ago. Their anxiety that even in the afterlife the authorities might still subject them to peremptory demands for labor tells us something significant about their experience of being ruled by a state.

PART THREE

THE EURASIAN LANDMASS

CHAPTER 7

THE ANCIENT NEAR EAST

I. The Life and Death of the World's Oldest Civilization

In the last few chapters we looked at the continents one—or even two—at a time. We now come to a landmass far larger than any of them, Eurasia. Because of its great size, and its even more salient role in history, we need to take it region by region. Conventionally it is divided into the two adjoining continents of Europe and Asia. But for most of our purposes this familiar division is unhelpful. Instead, we should probably start with a crude distinction between the cold regions to the north and the temperate regions to the south. For the present we can think of the cold north as offstage. People lived there, things happened there, and this activity could have drastic repercussions farther south, but the north was not the site of the development of the classic Eurasian civilizations. The onstage regions to the south stretch in a rough band across Eurasia, with the Near East in the middle. To the east are India and China, and to the west are the Mediterranean world and—partly overlapping with it—western Europe. The views to the east and west are far from symmetrical, but we can leave this asymmetry aside while we concentrate on the Near East.

The primary geological event in the making of the Near East was a collision between two continental plates. This collision, an event of the last fifty million years, joined Eurasia to the northeastern edge of what was then a greater Africa. From this to the familiar outline of the region today was a matter of only two steps. The first was a real geological event, the opening of the Gulf of Aden and the Red Sea over the last twenty million years or so; this partially separated the Near East from the diminished Africa we know today (for reasons of convenience, I am excluding Egypt from the Near East). The second step was just a superficial change, the formation of the Persian Gulf as a result of the high sea levels of the Holocene. In all this the most significant event was undoubtedly the collision of the two plates. The join runs from the northeastern corner of the Mediterranean near the island of Cyprus to the Gulf of Oman on the edge of the Indian Ocean. This line is of more than geological interest: it divides the Near East into two parts with significantly different characteristics.

The southern Near East consists of Arabia and the Fertile Crescent, the latter made up of geographical Syria (the eastern seaboard of the Mediterranean and its hinterland) and Mesopotamia (roughly the modern Iraq). Like the African plate of which it was originally part, this region is predominantly flat. Where it is mountainous, this is largely a consequence of the rifting system that extends into the southern Near East from East Africa. Since the Red Sea and the Gulf of Aden are branches of this system, it is no surprise that Yemen, the adjoining part of Arabia, is mountainous in the same fashion as the Ethiopian highlands. Farther north yet another branch of the system runs up the middle of Syria, where it is associated with two parallel mountain ranges. But in general the southern Near East is a land of plains.

The northern Near East, by contrast, is the region that bears the scars of the intercontinental collision. The result is less impressive than the Himalayas, but Iran and Anatolia (roughly modern Turkey) are generally mountainous in character. The main exceptions are the relatively flat plateaus found in the center of each.

To understand the Near Eastern climate, we should first note how the region is sandwiched between two major landmasses, Africa and northern Eurasia. On the African side this proximity exposes the Near East to hot air from the Sahara, of which the Arabian desert is in effect an eastward extension. On the Eurasian side the equivalent is cold air from Siberia. Since these are among the world's most effective heating and cooling systems, they do much to explain the extremes of temperature to which the Near East is subject. But there is one thing that continental airmasses, whether hot or cold, tend to have in common: they are dry.

This means that if the Near East is to enjoy a decent supply of rainfall, it has to obtain it from the adjoining seas. Here we can think of the region as sandwiched between the Indian Ocean and the Mediterranean. Of the two, the Indian Ocean is at first sight the more promising. It is vast, it has very high evaporation rates, and it generates enormous amounts of rainfall. But owing to the direction of the prevailing winds, the Near East derives very little benefit from its proximity to this ocean—which is why Arabia is a desert and not a tropical jungle. The Near East is thus left with the Mediterranean, a far less generous source of rainfall, and even then only in winter. The main factors enabling a region to draw on this source are proximity and the possession of mountains. The effect of these is to establish a marked aridity gradient across the Near East: overall it is wettest in the northwest, and driest in the southeast (see map 5). Thus only about 3 percent of the surface area of Turkey is desert, whereas for Saudi Arabia the figure is more like 97 percent (things were better a few thousand years ago). This makes Arabia so arid that in this chapter most of it will be offstage—though it will make a dramatic appearance closely linked to its aridity in a later chapter.

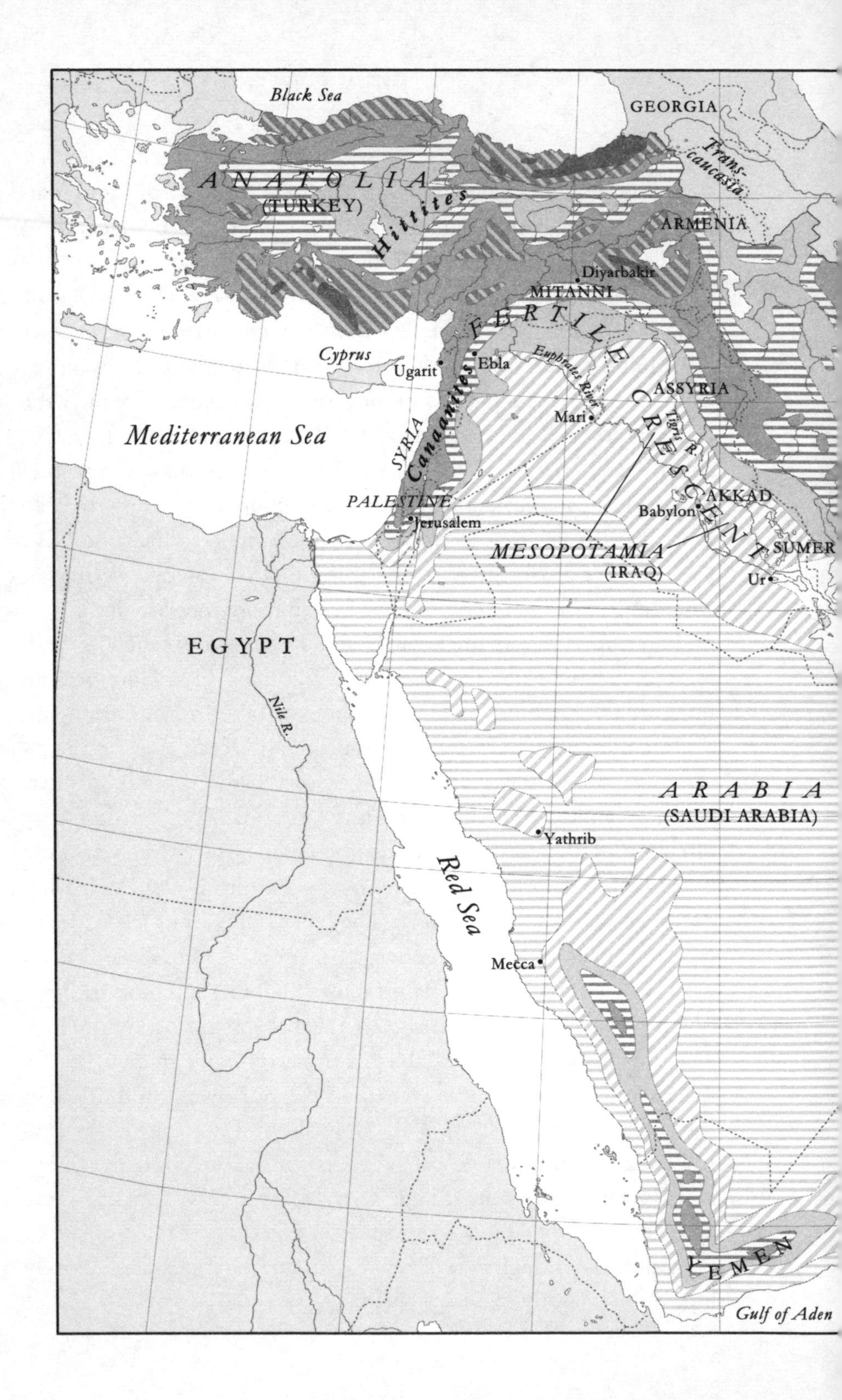

Black Sea
GEORGIA
Trans-caucasia
ANATOLIA
(TURKEY)
Hittites
ARMENIA
Diyarbakir
MITANNI
FERTILE CRESCENT
Cyprus
Ugarit
Ebla
Euphrates River
ASSYRIA
Mari
Tigris R.
Mediterranean Sea
SYRIA
Canaanites
PALESTINE
Jerusalem
AKKAD
Babylon
MESOPOTAMIA
(IRAQ)
SUMER
Ur
EGYPT
Nile R.
ARABIA
(SAUDI ARABIA)
Yathrib
Red Sea
Mecca
YEMEN
Gulf of Aden

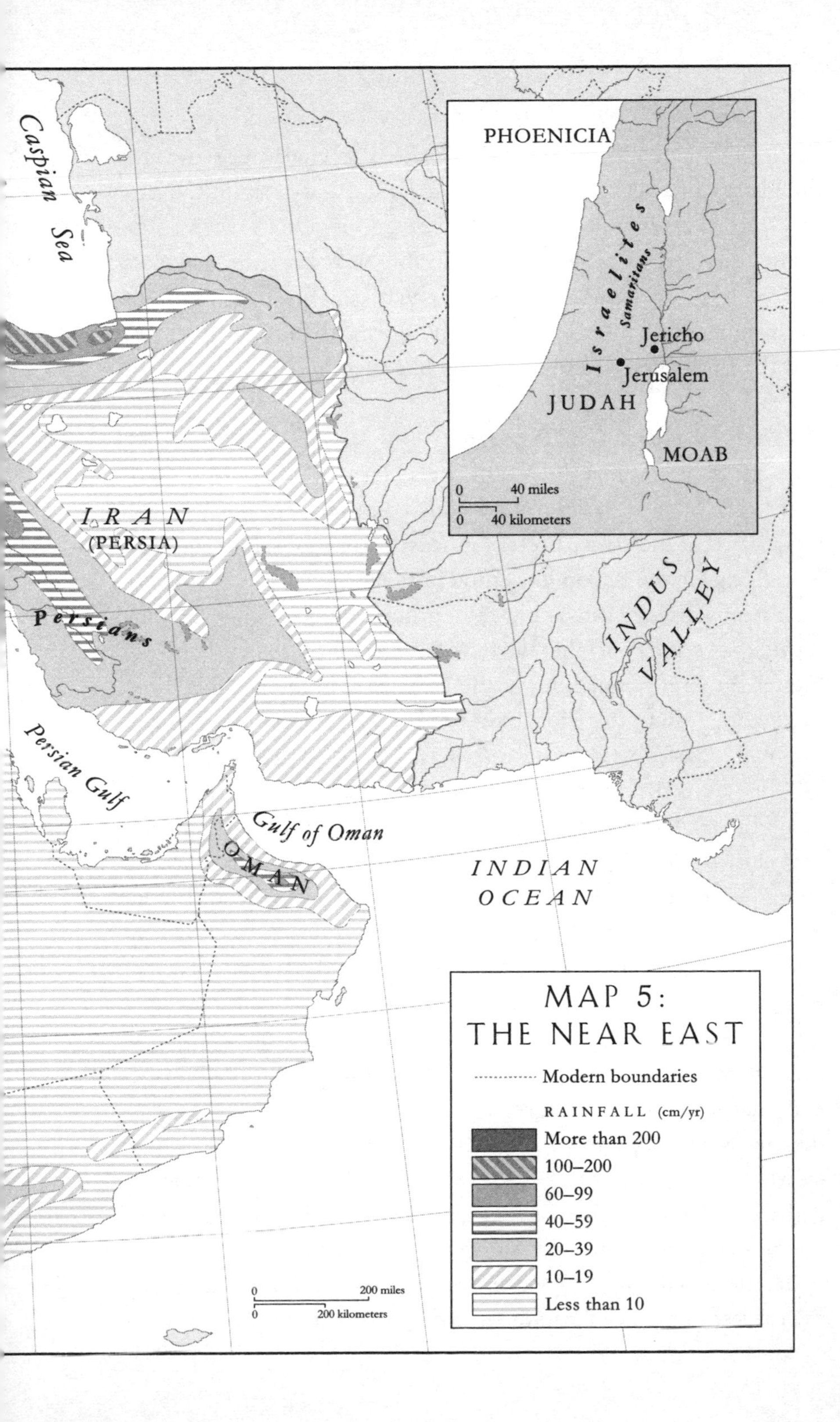
Caspian Sea
IRAN
(PERSIA)
Persians
Persian Gulf
Gulf of Oman
OMAN
INDIAN OCEAN
INDUS VALLEY
PHOENICIA
Israelites
Samaritans
Jericho
Jerusalem
JUDAH
MOAB
0 40 miles
0 40 kilometers
0 200 miles
0 200 kilometers
MAP 5:
THE NEAR EAST
Modern boundaries
RAINFALL (cm/yr)
More than 200
100–200
60–99
40–59
20–39
10–19
Less than 10

One major qualification to this picture is of vital importance for the history of civilization in the Near East. As we saw in the case of Egypt, it is possible for an otherwise arid region to receive substantial supplies of water by river. In the Near East the region so blessed is southern Mesopotamia, which benefits from the relatively abundant winter rainfall of the north, thanks to the Tigris and the Euphrates. There is nothing like them in Arabia.

Near Eastern Civilization

It stands to reason that the Near East is likely to have been the earliest region outside Africa to be colonized by modern humans, and it is clear that they were present at least forty thousand years ago. Yet what gave the region its central importance in history was not their relatively early arrival but rather their precocity in developing farming—so much so that we have no record of the existence of hunter-gatherer populations in the Near East in the last few thousand years. Add to this the rapidity with which the advent of farming transformed Near Eastern society. Figure 14 conveys a sense of this transformation: it shows what part of an Anatolian Neolithic settlement with a population of several thousand may have looked like in the seventh millennium B.C. There are no streets; entry to a building was through a hole in the roof. By our standards, let alone those of hunter-gatherers, this is an extraordinarily claustrophobic way to live. It may indeed have seemed that way to the inhabitants of the settlement at the time, but it had one distinct advantage: it presented a blank wall to the outside world. Long before this the inhabitants of Neolithic Jericho had apparently felt it necessary to surround themselves with fortifications, thereby becoming one of the earliest gated communities. It was clearly very important to keep people out. The life of the hunter-gatherers of the day may have been brutish and short, but that of the Near Eastern cultivators showed a new nastiness.

This precocity of the Near East also does much to explain why the earliest known civilization should have emerged there. It does

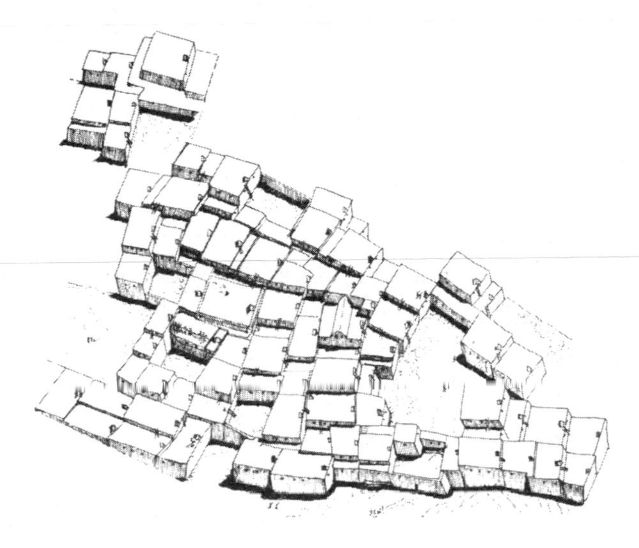

Fig. 14: A tentative reconstruction of houses at the Neolithic settlement of Çatal Hüyük.

not explain why this development should have taken place in southern Mesopotamia. That region, like Egypt, was a late starter in the Neolithic world. Again like Egypt, it was made up of a stark combination of river and desert. This environment gave to these lands an unusual combination of advantage and disadvantage. The advantage was that irrigation could generate a more substantial and accessible agricultural surplus than rainfall agriculture. The disadvantage was that other critical resources were lacking, and could be procured only through long-distance trade. In the case of southern Mesopotamia, these included materials as basic as timber, metals, and even stone. Thus in the fifth millennium B.C. the hapless cultivators were using clay sickles; later these were replaced, first with flint and then with copper—both of which had to be imported. In one way or another the earliest civilizations, and that of Mesopotamia in particular, can be seen as social and institutional responses to such imbalances. A noteworthy feature of the archaeological record in this connection is the greatly expanded commercial horizons of southern Mesopotamia in the centuries leading up to the emergence of civilization.

It was the Sumerians who played the key role in this emergence. They lived in a political framework quite unlike that of ancient Egypt: instead of forming a unified kingdom, they were divided into some thirty city-states. After a millennium or so the Sumerians themselves faded away; they were no doubt assimilated by their neighbors, as happens often enough in history, though we have no idea what caused it in this particular instance. But by then their civilization had been adopted by other peoples, notably the Akkadians to their northwest. The tradition the Sumerians initiated was to remain the culture of Mesopotamia down to the second half of the first millennium B.C.

Since Mesopotamian civilization, like that of Egypt, was fated to die out, none of its literary remains have been transmitted to us directly. As with ancient Egypt, we derive most of what we know from chance finds or archaeology. In the Egyptian case our main debt is to the desert conditions that have preserved large quantities of ancient papyrus. In the Mesopotamian case papyrus would not have survived, and it is our good fortune that the prime material used by cuneiform scribes was clay. Once fired, whether deliberately or by accident, clay tablets can survive as well as pottery can. The result is to give us a fragmentary, but very diverse, body of ancient Mesopotamian sources. Much of it is the product of more or less meticulous administration, but there is also material of many other kinds, including one textual tradition of considerable literary appeal, the Gilgamesh epic.

In political terms this civilization never did develop a tradition of a unitary Mesopotamian state. There was, however, a tendency for the scale of political organization to increase over time. In the third millennium B.C. city-states were the norm, while an imperial episode in the twenty-fourth and twenty-third centuries was the exception; in the first millennium B.C. the city-states were gone, and empires like those of the Assyrians and Babylonians had become the norm.

As was mentioned in the preceding chapter, Mesopotamian civilization proved far more exportable than that of ancient Egypt. Indeed, it is likely that Mesopotamian influences played some part in the formation of the nearby civilizations of Egypt and the Indus Valley. But what concerns us here is a more straightforward process, the undisguised adoption by other peoples of the literary culture created by the Sumerians. Already in the third millennium the neighboring Akkadians had done this, as had the people of Ebla in northern Syria. An example of the same phenomenon from the second millennium is provided by the Hittites of Anatolia. Typically what such peoples did, at the level of literary culture, was two things. They adopted the Sumerian literary tradition, with the result that Sumerian became the world's first classical language, still cultivated by the educated long after it had died out as a language of everyday life even in its own land. And at the same time they took steps to adapt the cuneiform script to the writing of their own languages, thus giving us our earliest direct knowledge of the Semitic and Indo-European language families—Akkadian for Semitic, and Hittite for Indo-European.

As we will see in the third section of this chapter, a remarkable feature of this literary tradition was its unprecedented awareness of its own continuity. But by the first millennium B.C., in Mesopotamia as in Egypt, this continuity was beginning to have an archaic ring to it. The world had changed.

This change was most palpable at the level of political power. Like Egypt, Mesopotamia was exposed to the risk of foreign invasion, and thanks to its much less sheltered location the danger was often more pressing. It suffered a significant intrusion in the third millennium B.C., and matters were much worse in the second. Sometimes the intruders were the hill tribes of western Iran, as with the Gutians of the third millennium and the Kassites of the second. Sometimes they were the pastoralists of the southern Near East, as with the Amorites around 2000 B.C. and the Arameans a millennium later. But something all these invasions, large and small, had in common was that the newcomers as yet posed no threat to the cultural continuity of Mesopotamian civilization. The Amorites, for example, are described as a people who dwell in tents and have no knowledge of houses or grain. If such people successfully established their power in Mesopotamia, they soon found that they needed the appurtenances of civilization; and since they lacked one of their own, they adopted that of their subjects.

In the first millennium B.C. it might have seemed as if the pattern would continue as before. In the sixth century Mesopotamia was again invaded from Iran, this time by the Persians, who went on to devise a cuneiform script for their own language, and to use it in their royal inscriptions. But there was a difference: Mesopotamia was now only one of several cultural provinces of a very large empire. There was worse to come. In the fourth century the Greeks overthrew the Persians, and this time the invaders brought with them an urban civilization of their own.

There was also a subtler subversion at work that had its origin in Syria. Unlike neighboring Egypt and Mesopotamia, Syria was neither embellished nor weighed down by a third-millennium cultural tradition. Instead, the keynote of Syrian cultural history was instability and innovation. In the third millennium Ebla shows us a typical local adoption of the Sumerian literary tradition. Yet by the middle of the second millennium Syria had broken with the cumbersome script of Mesopotamia and adopted the alphabet. This new technology proved immensely attractive and spread far and wide. Thus it was received in Yemen, even though this remote region had shown no interest in earlier writing systems. But what concerns us here is the use of the alphabet to write Aramaic, the language that in the course of the first millennium B.C. displaced the older Semitic tongues of Syria and Mesopotamia in daily life.

Unfortunately for us, alphabetic writing in Mesopotamia was normally done on perishable materials, not on clay, with the result that little of it has survived. But it seems that even in Mesopotamia the old cuneiform tradition was losing ground to the new alphabetic culture. And it was Aramaic, not Akkadian, that was to be used as the primary written language of administration in the Persian Empire.

From the Persians to the Muslims

There is a millennium of Near Eastern history between the formation of the Persian Empire in the sixth century B.C. and the rise of Islam in the seventh century A.D. Like much else in this book, it will be shortchanged; but we can at least identify its major themes.

One is the political dominance throughout the period of peoples coming broadly from the north. Often they originated in Iran, as with the Persian Empire from the sixth to the fourth century B.C., the Parthian empire from the third century B.C. to the third century A.D., and a second Persian Empire from the third to the seventh century A.D. But they might also come from outside the Near East altogether, as with the Greeks from the fourth century B.C. and the Romans from the second century B.C. Of all these northern peoples, it was the Greeks whose cultural impact on the Near East was the greatest at the time.

The second theme is the eventual displacement of the ancient polytheism of the region by a monotheist faith, Christianity. This religion was an offshoot of a monotheist cult developed by a minor, but innovative, Near Eastern people, the Israelites. The emergence of Israelite monotheism is accordingly the subject of the next section of this chapter. But the fact is that as long as this phenomenon was confined to the Israelites and their ethnic heirs, the Jews and Samaritans, it was of no great historical importance. What made an idiosyncratic ethnic tradition a major ingredient of world history was its subsequent metamorphosis into two world religions, Christianity and Islam; their combined adherents currently amount to about half the world's population. We will inevitably come back to both of them in later chapters.

The final theme, unsurprisingly, is the demise of the ancient civilization of Mesopotamia. Like that of Egypt, it survived the end of native rule by some centuries. The last datable cuneiform tablet we possess is an astronomical almanac relating to A.D. 74–75. We cannot pinpoint the final extinction of Mesopotamian civilization, but by the third century A.D. it must have been effectively dead. In that century Mani, a Mesopotamian religious reformer, created a grand synthesis of the religious traditions of his world. Christianity and Buddhism were recognized, as was the Zoroastrian tradition of Iran. But the religious beliefs of ancient Mesopotamia found no place in Mani's new religion. It was as if they had never been.

II. Downsizing the Pantheon

According to Muslim tradition, some pagan Arabs of Mecca once expressed their sense of the absurdity of the mission of the prophet Muhammad by asking him, "Muhammad, do you want to turn the gods into one god? What a weird idea!" Whether it is weird or not, the idea has achieved wide currency and shaped the course of history. It emerged from the polytheism of the ancient Near East, though just why it did so is hard to say.

How old is polytheism? Archaeology has unearthed abundant indications of the religious proclivities of prehistoric humans. In the Upper Palaeolithic there are well-attested cases of burials with grave goods. In the Neolithic the evidence is much richer. There are, for example, spaces clearly set aside for religious activity; several of the units shown in figure 14 have been identified from their contents as shrines. But the archaeology of preliterate peoples can never tell us what they actually believed. The invention of writing breaks this silence: for the first time we can be certain that we are dealing with gods—and, better still, we can learn something of the myths associated with them.

From the beginning this record shows that the peoples of the ancient Near East believed in many gods. This is hardly surprising. Anyone who thinks that the world is run by divine will has to take account of the fact that the results, though not chaotic, look a bit disorganized; this suggests the involvement of more than one god, and that the gods are sometimes at cross-purposes. Moreover, the idea of a single god taking care of all aspects of the life of the universe seems rather impractical. For example, it is not obvious how the roles of a war god and a goddess of love could be combined felicitously. People living in a reasonably complex society would surely prefer to think in terms of a division of divine labor. All this, of course, sets out from an assumption: that the gods are rather like us. But who else would they be like?

The narrative of the Babylonian creation epic ends with a divine banquet attended by the fifty great gods and numerous others. They have gathered to celebrate, but there is also a political agenda: the gods swear an oath confirming that Marduk, the god of Babylon, is to exercise the kingship of the gods. They had in fact agreed to this rather burdensome arrangement at a previous banquet. On that occasion a generous supply of liquor had certainly helped to overcome any reservations they might have had. But there was a more pressing reason for their submission: they were in mortal danger, and Marduk was their only hope of deliverance. The problem was an ugly divine civil war that had broken out after the persistent rowdiness of the younger gods had alienated their primeval forebears. In the course of the struggle these younger gods became thoroughly demoralized. It was Marduk's ruthless and successful prosecution of the war that cemented the new regime, to such an extent that the gods then spent a year shoveling and making bricks to build the city of Babylon for him (it was there that the celebratory banquet took place). This, at least, was how the Babylonians told the story. When the Assyrians adopted the epic, they customized it by transferring the leading role to their own national god, Assur.

Matters were not so different in northern Syria, to judge by fourteenth-century texts from the city of Ugarit. The language of these texts is a form of Canaanite written in an alphabetic script; but fortunately for us they were inscribed on clay tablets. Here too the gods assemble, eat, and drink. In the Ugaritic pantheon the king of the gods is El, but he does not rule with the same authority as Marduk. Brutal conflicts unfold between Baal, the rain god, and two of his peers: Yamm, the god of the sea, and Mot, the god of the dry season. In the course of these brawlings El is not always treated with the respect that might be thought his due. Baal at one point stands in the divine assembly and spits; on another occasion a war goddess who is in love with Baal puts pressure on El by threatening to beat him up. (The sex between the divine lovers, incidentally, is as vividly described as the violence.) In each of these conflicts Baal proves to be the victor. In recognition of the status he has now achieved, El at last addresses a long-standing complaint of Baal's: an appropriate residence is built for him. This time, however, the hard work is left to the craftsman god.

It was a few centuries after this and a few hundred miles to the south that monotheism, or something close to it, emerged among the Israelites. In tracing this new departure, we are largely dependent on the Bible. This book is a collection of writings, mostly in Hebrew, which like Ugaritic is a form of Canaanite. Again the script is alphabetic, but it was written on perishable materials; we owe the survival of these writings not to archaeological good luck but to continuous literary transmission. The Bible, in fact, is the only body of ancient Near Eastern texts that has reached us by such a route, and we should be duly grateful for it. Yet this mode of transmission has problems of its own. Unlike an assortment of clay tablets from an archaeological site, such a corpus is likely to be the product of a prolonged process of later winnowing and editing. In the case of the Bible the key roles in this process must have been played by committed monotheists, with the result that allowances have to be made for the likely tendentiousness of the Biblical account of the Israelite past. So it may be better to try to reconstruct the emergence of monotheism by starting at the end and working back.

In 587 B.C. the Babylonian ruler Nebuchadnezzar destroyed Jerusalem and deported its Israelite or, more precisely, Jewish population to Mesopotamia. In a part of the book of Isaiah written a few decades later among the Jewish exiles in Babylon, there are several divine affirmations of the form "I am Yahweh and there is no other," "I am God and there is no other" (e.g., Isaiah 45:18, 22). What these assert—and they seem to be more than just rhetoric—is that other gods do not exist. At first sight this is counterintuitive: political and military disaster is strong prima facie evidence that other people's gods are alive and well. But on the basis of this and other texts, we can be reasonably confident that by the sixth century B.C. a full-fledged monotheism had appeared.

It is also clear that this unqualified monotheism is not the standard message of the Bible. In the words of the Ten Commandments, God tells the Israelites that "you shall have no other gods before me" (Exodus 20:3), not that other gods do not exist. In fact, numerous passages speak plainly of the existence of other gods. For example, in the story of the exodus from Egypt, God says, "on all the gods of Egypt I will execute judgments" (Exodus 12:12). We even see him sitting in the assembly of the gods: "in the midst of the gods he holds judgment"—though the gods are then told that they will die like mortals (Psalms 82:1, 7). In this conception the relationship of the Israelites to God is analogous to a marriage: if you enter into a monogamous marriage, you renounce all other potential partners, but nobody expects you to deny their existence. What we have here is not exactly monotheism but rather monolatry, the worship of one god to the exclusion of others.

How early in Israelite history did this monolatry emerge? As far as the Bible is concerned, it had been there as far back as Abraham, which takes us to pre-Israelite times. But the Bible also tells us that the Israelites kept backsliding. Solomon, who in the tenth century built a temple for the national god "to dwell in forever" (1 Kings 8:13), also established cults of other gods for his foreign wives, and apparently participated in them himself (1 Kings 11:5–8). The prophet Jeremiah in the early sixth century had an exchange with Jews living in Egypt who worshiped the "queen of heaven" (Jeremiah 44:15–19). They told him in no uncertain terms that they would continue to worship her "just as we and our ancestors, our kings and our officials, used to do in the towns of Judah and in the streets of Jerusalem." The women were particularly insistent—appropriately, since the cult of a goddess was at stake. None of those involved showed the slightest sense that they or their ancestors had done anything improper. Even as a principle, then, it seems that strict monolatry was far from universally espoused in ancient Israel.

What may well have been both old and widespread was a strong—though not exclusive—focus on the national god. In contrast to monolatry and monotheism, a special relationship of this kind would not have been out of place in the ancient Near East. A ninth-century inscription set up by a king of Moab, an eastern neighbor of the Israelites, speaks of Chemosh, the national god, in a tone reminiscent of the older parts of the Bible. Meanwhile, the Bible refers to the Moabites as "the people of Chemosh" (Numbers 21:29), just as it calls the Israelites "the people of Yahweh" (Judges 5:11). Even in Mesopotamia, as we saw, the Babylonians put Marduk in the limelight, while the Assyrians did the same for Assur. This relationship of a people to its national god provides a plausible starting point for the evolution of monolatry and, eventually, monotheism. But the plain fact is that we do not know why so familiar a feature of ancient Near Eastern religion evolved in such an unusual direction in the Israelite case.

We should not, however, overstate the radical nature of the evolution. The monotheist tradition that issued from ancient Israel is in one respect very conservative. It takes for granted that God is a god, as in the Muslim profession of faith: "There is no god but God." Monotheism thus shares with ancient Near Eastern polytheism the premise that there is such a thing as a god; the disagreement is limited to the question of how many there are. It was not in Israel but in Egypt that this unthinking continuity with the polytheist heritage was challenged. There in the fourteenth century B.C. King Akhenaten instituted an exclusive cult of the solar disk. We do not know whether he denied the existence of the traditional Egyptian gods or simply spurned them; we do know that as his reign wore on he became less and less inclined to apply to the object of his cult the old polytheistic word "god." What he worshiped, by implication, was something different. It seems, then, that he came to reject not just the particular gods of his country but the category itself. Some fifteen hundred years after Akhenaten the Egyptians finally abandoned their traditional gods and adopted a zealous Christian monotheism. Ironically, the word they then applied to the Christian god was the same polytheistic term that Akhenaten had discarded.

Akhenaten's cult lasted hardly longer than his reign, but monotheism in its various forms—Jewish, Christian, Muslim, and others—is still very much with us today. Why it should have been so successful is a long story that we will touch on from time to time in later chapters, though without ever quite coming up with an answer. But if we are to pick out a single aspect of monotheism in this connection, it might be less what monotheism embraces than what it rejects: other gods and the people who worship them. It is in the nature of monotheism to pick a quarrel. This exclusive spirit is normally foreign to polytheism, and it lies behind a variety of historical phenomena, from the remarkable survival of the originators of monotheism as an ethnic group to the specter of monotheist terrorism that haunts the world at the beginning of the twenty-first century.

III. Archaism

Human societies do not appear out of nothing; any society that exists today is a continuation of some earlier one. Naturally the picture is complicated by the fact that societies not only change but also fuse and split. Thus a complete genealogy of an existing society would not go back to the beginning in a single line. But go back it would—we have no reason to doubt that human society has existed continuously as long as the human race.

Of course no existing society can give a remotely credible account of itself on such a timescale. The collective memories of nonliterate peoples, who for most of the human past were the only peoples there were, do not seem to go back more than a few centuries at most. Remembering things is an effort, and the effort is hardly worth making when the things in question are no longer relevant. The Incas offer a case in point: beyond their own dynastic history, they had only a myth of origin. Among the Jie of East Africa, the fact that the successive age-groups had names provided the tribe with a way of tying events into a comparable chronological framework. On this basis one modern historian has traced back their collective memory to the early eighteenth century. But within a few centuries, as we probe the past of a nonliterate people, we reach a point at which there is only myth or amnesia.

This is doubtless how matters were in Mesopotamia before the development of writing, and would have continued thereafter without it. As in Egypt, however, writing eventually made possible an enormous extension of the horizon of collective memory. As we saw in the preceding chapter, Manetho in the early third century B.C. could say who had ruled Egypt since around 3000 B.C. His Babylonian contemporary Berosus, who likewise wrote about the past of his native land in Greek, was probably able to do the same for Babylon, though our very limited knowledge of what he wrote makes it impossible to confirm this.

One result of this expanded horizon was that people could go back into the past—*their* past—and retrieve aspects of it that had long ago fallen into disuse.

Egyptian culture of the seventh and sixth centuries B.C. is marked by a phenomenon known to Egyptologists as archaism. To a far greater extent than in other periods of Egyptian history, kings and members of the elite were not content just to continue their cultural tradition as they had received it. Instead, the art and architecture of their funerary monuments hark back to much earlier models, some as old as the third millennium B.C.; and the language of the associated inscriptions is for the most part a classical Egyptian very different from the language spoken and written in ordinary life. We do not know exactly what the archaizers thought they were doing, because they do not tell us. Were they setting out to restore a glorious past in the present? Or were they just picking and choosing what took their fancy in the ruins that surrounded them? Either way, the distant past was now a resource for them.

What we find in Mesopotamia in the same period is more interesting, precisely because there people occasionally paused to explain themselves.

Restoration was clearly in the air in sixth-century Babylon. The last Babylonian ruler, Nabonidus, has left us a remarkable account of his revival of an institution that had been in disuse for over a millennium. In the twenty-fourth century B.C. the imperial ruler Sargon had installed his daughter (incidentally the world's first known female author) in the office of high priestess and human consort of the moon god Nanna in his temple at Ur. Whether or not Sargon was the first to do this, the practice was followed by other rulers in Mesopotamia down to the eighteenth century; thereafter we hear no more of it until Nabonidus revived it. His claim was that the moon god had communicated his need for a high priestess and that divination then established that she had to be the king's own daughter. But at this point Nabonidus faced a problem. To revive an ancient practice requires more than just a general idea of what it once involved, and in this case, as Nabonidus pointed out, the office had been forgotten for a very long time. Fortunately, he tells us, he was able to uncover relevant materials dating from the reign of a twelfth-century king (did he really find them, or did he only pretend?). Thus he was able to do things just as they had been done in ancient times.

His stated motive for all this was his concern for the sanctuaries of the great gods. We can accept this pious profession, albeit with the qualification that Nabonidus had good reason to make a conspicuous display of his concern: he had come to the throne in a manner that did not bear scrutiny, and was scrambling for legitimacy. What is interesting is that the way in which he manifested his concern was an extravaganza of archaism.

Nabonidus was also an indefatigable excavator, and in this he was not alone among the later Babylonian kings. The context was again his pious concern for the sanctuaries of the great gods: he sought to rebuild ancient temples that had fallen into ruin or disappeared altogether. It was believed at the time that such a restoration would not be valid unless the new temple was erected on the exact site of the old. It was therefore necessary to find the old foundations. But this could still leave a problem: What if one had the wrong foundations, as could easily happen if the precise location of the ancient temple had been forgotten over the centuries? For this reason great efforts were made in the course of these excavations to find the foundation inscriptions left by earlier kings. In one case Nebuchadnezzar had rebuilt a temple in the early sixth century on a site where he had found an inscription left by an earlier restorer. Later Nabonidus discovered a yet earlier inscription in a different place and rebuilt the temple accordingly.

These Babylonian excavations were not archaeology in our sense; what inspired them was a contemporary religious concern, not an academic interest in the past. But it might be going too far to conclude that Babylonian kings had no interest whatever in antiquities for their own sake. Figure 15 shows a statue of the time

Fig. 15: A statue originally from Mari, around 2000 B.C.

of a certain Puzur-Ishtar, who was governor (in effect ruler) of the city of Mari, some way up the Euphrates from Babylon, around 2000 B.C. It is a fine museum piece. In fact, it may already have been in a museum in the sixth century B.C.; it was found in a collection of antiquities that seems to have been brought together by Nebuchadnezzar in his palace in Babylon, and added to by his successors.

The new ability to retrieve the past that arises with the creation and survival of literary heritages is more important in human affairs than these first Mesopotamian examples would suggest. Heading back into your distant past can be an ingenious way to circumvent your present. Of course, if you seek something different from what you currently have, you can always borrow from some other culture. But the advantage of your distant past is that, though very different from your present, it is nevertheless your own. Hence the remarkable power in human history of renaissances, reformations, and fundamentalisms.

CHAPTER 8

INDIA

I. Why India Was Not Just a Subcontinent

Like the southern Near East, India is a part of the old supercontinent of Gondwana that has joined Eurasia. But the sequence of events was different: India separated from the rest of Gondwana over a hundred million years ago, and until it collided with Eurasia, about fifty million years ago, it led the existence of an island. This distinctive history is reflected in the single most obvious difference between the two regions: the Indian subcontinent is a peninsula, not a land bridge. There are nevertheless some broad physical similarities. Like the southern Near East, India overall is relatively flat. Inland from the coastal plains of the south there are mountain ranges arising from the rifting that brought the region into being, and between them are the highlands of the interior. But there is nothing in the south to compare in altitude to the combination of mountains and plateau by which India is closed off to the north. Much as in the northern Near East, this massive uplifting is the product not of rifting but of collision. Yet the scale of the uplifting dwarfs anything the Near East has to show; indeed, there may have been nothing on earth to match it in the last half billion years. By contrast, it is between the northern mountains and the southern highlands that India is at its lowest and flattest, with alluvial plains comparable to those of Mesopotamia.

One of the most significant differences between India and the Near East is climatic: India gets much more rain (map 6). Given the high terrain to the north and the open ocean to the south, this makes intuitive sense; and it is in fact this combination of physical features that generates the monsoon, the wet summer so alien to the Near East. But the distribution of summer rainfall in India is very uneven. The wettest regions are the coastal strip in the southwest and a large area in the northeast; by contrast, the southern highlands are a bit dry, and the northwest shares the aridity of the Near East, of which it is in effect a climatic extension. As a result, the two great rivers that rise in the northern mountains have somewhat different effects on the lands through which they flow. Both of them deliver a valuable agricultural resource, namely silt. But in the northeast the Ganges takes its water to a region of abundant rainfall designed by nature to be a jungle; whereas in the northwest the Indus—like the great rivers of the Near East—brings life to a land much of which would otherwise be desert.

For our purposes, it makes sense to divide the historical India into three major regions. The first is the south, with its coastal plains, modest mountain ranges, and highland plateau; it is by no means lacking in agricultural resources, but they are somewhat dispersed. The second region is the northeast, centered on the Ganges; this has an extraordinary concentration of rich agricultural land. Finally there is the northwest, today Pakistan, centered on the Indus; here irrigated agriculture is possible as in Egypt or Mesopotamia.

An important respect in which these three regions of India differed in earlier times was their potential for contact with the outside world, with all its costs and benefits. The most remote area was the south. Until the development of navigation on the open oceans, it had no direct contact with regions outside India, and within the subcontinent its physical character rendered it less accessible than the northern plains. The northeast differed from the south in that it was internally more accessible and adjoined other parts of Eurasia; but it was cut off from them by high mountains in the north and thick jungle to the east. The result was that its contacts with the wider world were routed mainly through the northwest. It was this northwestern region that, until recent centuries, served as the gateway to India—though it was by no means a commodious gateway, in that those who made use of it had to reckon with mountain and desert. But a glance at prehistory suggests that the role of the northwest in the making of India has been fundamental.

India's First Civilization

Archaeology shows that modern humans have been present in the subcontinent for at least thirty thousand years. It does not show by what route they arrived, but if they came from Africa it stands to reason that they would have entered from the northwest. Genetic evidence provides some support for this inference: it points to an early human expansion from the Horn of Africa along the southern coast of Eurasia, starting with Arabia and continuing to India and beyond. The route by which farming first entered India is more directly attested. Domesticated crops appear by around 6000 B.C. in the extreme northwest of the subcontinent, and only later are they found in the rest of India, reaching the south by the third millennium B.C. Toward the middle of the same millennium the Indus Valley civilization emerged. It is unlikely to have been an accident that this first Indian civilization took shape in the part of India that was closest to Mesopotamia and most similar to it—and never spread to the rest of the subcontinent.

Far less is known about the Indus Valley civilization than about that of Mesopotamia. We have no idea what the people in question

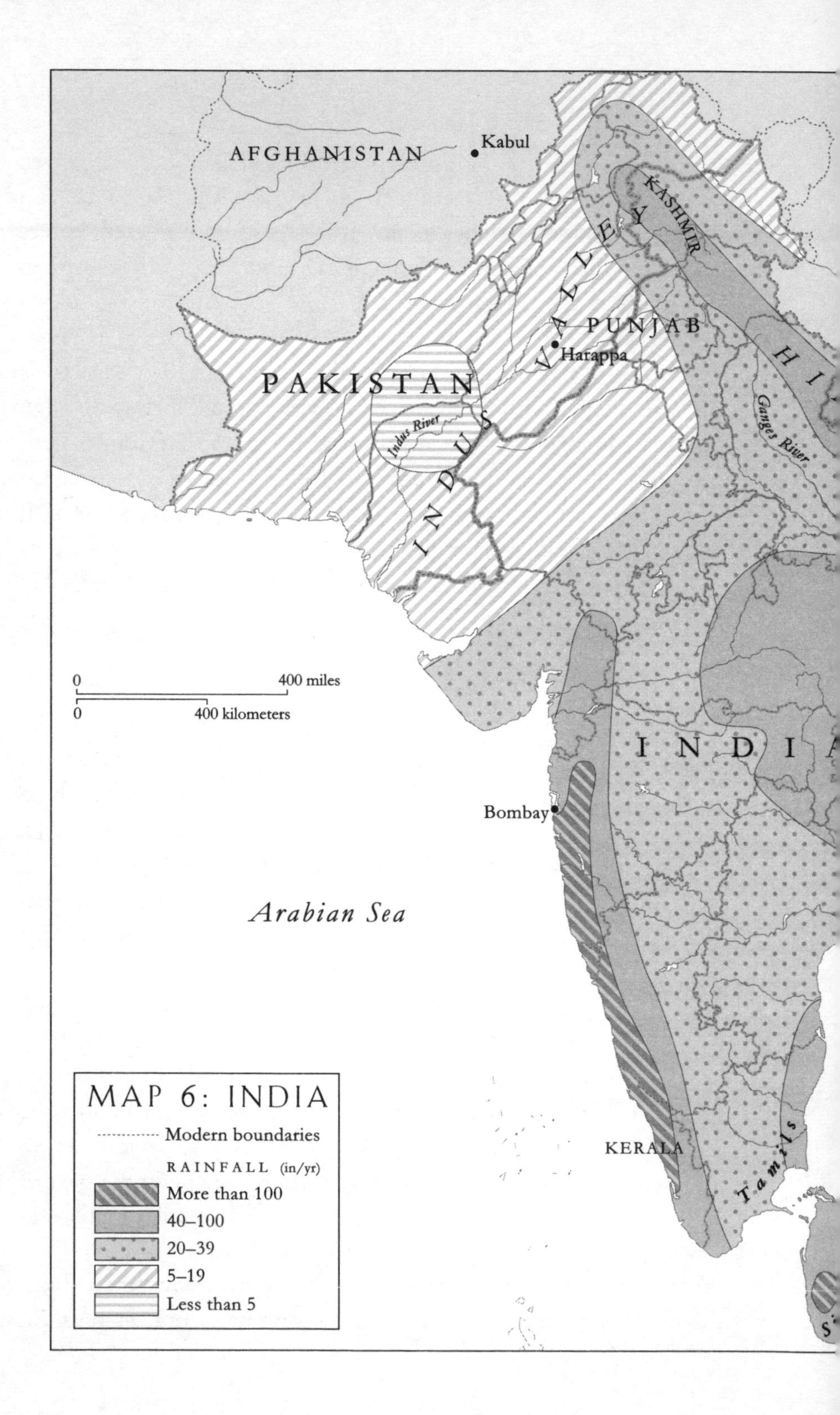

AFGHANISTAN
Kabul
KASHMIR
PUNJAB
Harappa
PAKISTAN
INDUS VALLEY
Indus River
Ganges River
0
400 miles
0
400 kilometers
INDIA
Bombay
Arabian Sea
KERALA
Tamils
MAP 6: INDIA
Modern boundaries
RAINFALL (in/yr)
More than 100
40–100
20–39
5–19
Less than 5

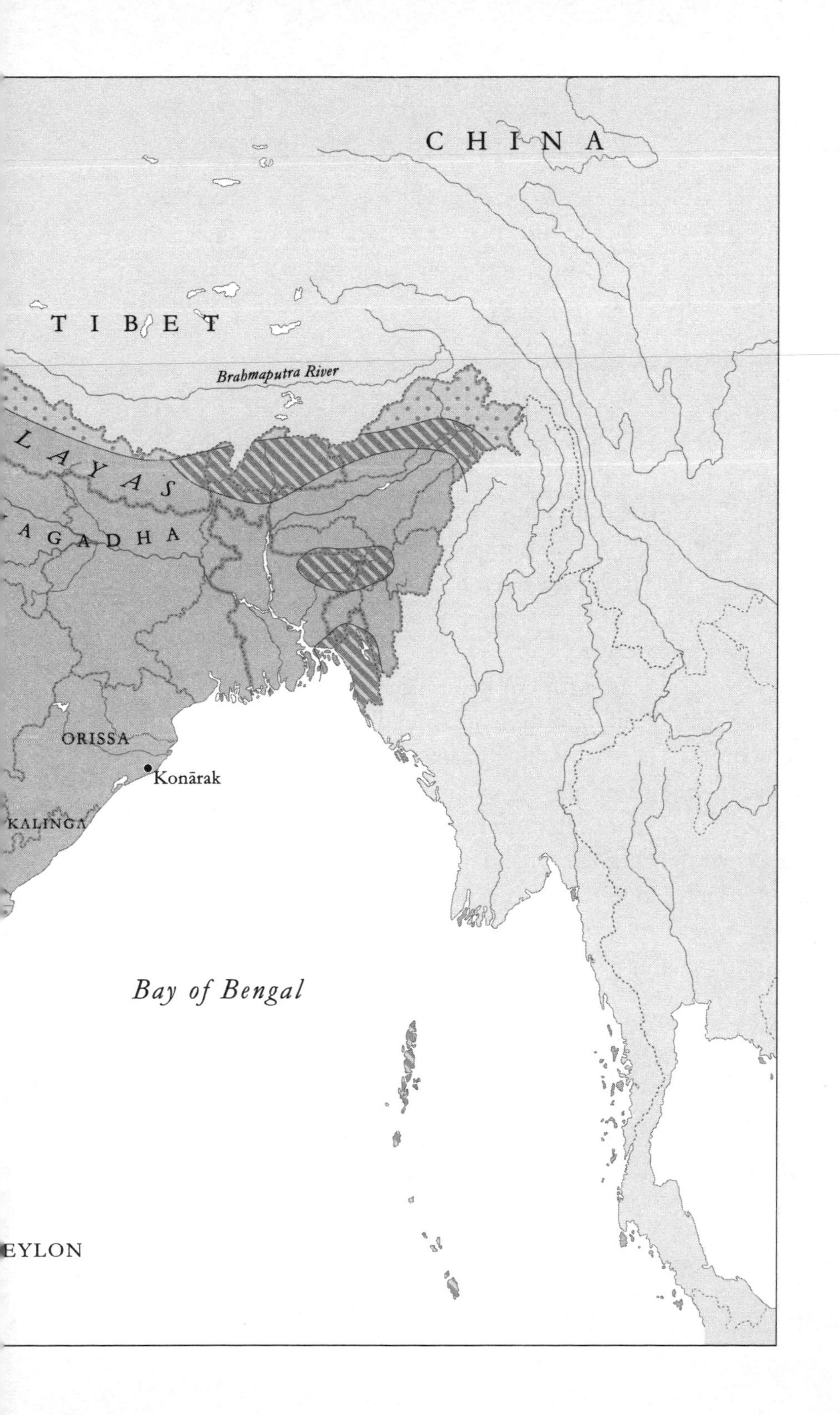

CHINA
TIBET
Brahmaputra River
LAYAS
AGADHA
ORISSA
Konārak
KALINGA
Bay of Bengal
EYLON

called themselves, or what kind of language they spoke; for convenience we often refer to them as Harappans, from the name of a major archaeological site that was once one of their cities. Archaeology shows, for example, that they had unusually regular town planning; but it leaves us to guess at the political structures that made this possible. There are two main reasons why our knowledge is so limited.

First, it is our bad luck that the written records left by this civilization are scanty. We have considerable numbers of inscribed stone seals; but the inscriptions are short, and cannot be deciphered. In marked contrast to the Near East, this region has left us no clay tablets or long inscriptions on stone. The obvious assumption is that the Harappans did most of their writing on perishable materials; if later Indian practice is anything to go by, they may have used palm leaves.

The second reason for our ignorance is as much their bad luck as ours: their civilization came to an end early in the second millennium B.C. There is nothing intrinsically mysterious about such an event—the surprise is, if anything, that early civilizations did not go extinct more often. In this instance the cause could have been environmental (for one thing, the Indus is a notably unstable river); or it could have been invasion. In any case, this early demise works against us in two ways. One is that later Indian tradition retains no memory of the Indus Valley civilization. This does not mean that the Harappans contributed nothing to the India of later times; some features of their culture do not go away, including perhaps a liking for the number sixteen manifested in their system of weights. But there is no equivalent to the historical information that the Chinese tradition preserves on the Shang dynasty. The other consequence of the early end of the civilization is that no outsiders have left us any account of it. There must have been people who could have done this, since there is clear evidence that the Indus Valley civilization was in commercial contact with Mesopotamia; in early Mesopotamian records we even hear of an interpreter. But it was not until the second half of the first millennium B.C. that ethnography came into its own, and by then it was too late.

The Aryans and Their Impact

At this point it is worth interrupting our narrative to consider the linguistic map of the subcontinent today. It is complicated on the northern fringes by the spillover of languages from adjoining regions, and in the hill country of east-central India by pockets of speakers of what are called the Munda languages, possibly an ancient intrusion from Southeast Asia. But otherwise the picture is remarkably simple. Two and only two language families occupy the great bulk of the subcontinent, Indo-Aryan in the north (e.g., Hindi) and Dravidian in the south (e.g., Tamil). The line separating them runs roughly from southwest to northeast; there is a clean break in the west, but scattered islands of Dravidian in the northeast.

The question we have to address is how this linguistic partition might have arisen. The obvious hypothesis is that the Dravidian languages were there before the speakers of Indo-Aryan arrived, and several things support this. One is that there is no consensus among linguists linking the Dravidian languages to any others outside the subcontinent. Another is that the scatter of Dravidian islands in the northeast invites interpretation as the residue of an originally wider presence (though the Dravidian-speakers of the far northwest may be migrants who arrived there in the last thousand years or so). Last but not least, the oldest form of Indo-Aryan we possess is already marked by the presence of some Dravidian loanwords; and since this was a language spoken in the northwest, these loanwords are a strong indication that this region, too, was once Dravidian-speaking (so it would be a fair guess that the Harappans spoke a Dravidian language). All this suggests that before the appearance of the Indo-Aryan-speakers, the Dravidian languages were spoken over most of India.

The Indo-Aryan languages, by contrast, are very well connected outside India. They form part of the Indo-European language family to which the Celtic, Germanic, Slavic, and several other groups of languages also belong. As we will see in chapter 12, these various groups are likely to have arisen from the breakup of a parent language in eastern Europe. But even without that reconstruction, the fact that only one branch of Indo-European is found in India, while all the others are located outside it, makes the presence of Indo-Aryan in India look like an intrusion.

Languages spread in two ways: people who speak them may move to new areas, and people who did not originally speak them may adopt them. Often both processes may be involved; the problem in such cases is to decide the relative weight of the two. With regard to India two things can be said with some confidence. One is that Dravidian-speakers would not have begun to adopt Indo-Aryan languages unless some Indo-Aryan-speakers had brought them to the subcontinent in the first place. The other is that for the incoming Indo-Aryan-speakers to have had the impact they did, they must have been either numerous, or powerful, or both. The idea that a significant number of people were involved has some support from a limited body of genetic evidence, which points to a marked affinity between high-caste Indians and Europeans—particularly eastern Europeans. The idea that a lot of power was involved is supported by the very fact that it is high-caste Indians who most display this affinity, and also by the finding that it is stronger on the male than on the female side.

Although we cannot hope to pinpoint the arrival of the Indo-Aryan-speakers in India, we can do something to give it a context. The first hard evidence of Indo-Aryan comes not from India but from the Near East. The place is the kingdom of Mitanni, in the region of what is now southeastern Turkey, and the date is the fifteenth and fourteenth centuries B.C. The spoken and written language of Mitanni was Hurrian, a local Near Eastern language; but Indo-Aryan words crop up in such suggestive contexts as royal names, names of gods, and the technical language of chariot driving. This looks like the residue of an Indo-Aryan military presence at some earlier date—except that these would be speakers of Indo-Aryan who never went to India.

In India itself in this period, and for a long time to come, we look in vain for what the Near East supplies in abundance—contemporary texts archaeologically preserved. Instead, we must work with two sources of diametrically opposed character. One is the mute remains of Harappan civilization and of post-Harappan cultures—an archaeological record without a soundtrack. The other is the oral tradition of the Indo-Aryan-speakers—a soundtrack that is hard to relate with confidence to the archaeological record.

Oral tradition has been a salient feature of Indian religious life for a very long time. In the mainstream Hindu tradition the Brahmins were the priests, and one of their key roles was to memorize and transmit to future generations enormous bodies of textual material related to the cult. The oldest and most prestigious of these texts were the four Vedas, transmitted in a number of recensions by the various Brahmin groups of the different parts of India. Of the four Vedas, the oldest was the Ṛgveda, a substantial collection of hymns addressed to the gods. Here, as in many aspects of Indian history, the chronology is vague, but scholarly consensus dates the composition of this material to the second half of the second millennium B.C. The geography is less of a problem: we are in the eastern Punjab, a part of the northwest where the various tributaries of the Indus have not yet joined to form a single river—just where we would expect to find people who had recently entered India through the mountain passes. The subsequent transmission of these hymns seems to have been extraordinarily faithful. One indication of this is the fact that these Bronze Age compositions have not been contaminated by the Iron Age, in which the great majority of those who transmitted them actually lived (by contrast, anachronistic references to iron have seeped into the portrayal of Bronze Age society in the Homeric epics). The language of the hymns is an archaic form of Sanskrit, the Indo-Aryan classical language of Indian civilization; the Ṛgveda thus provides our earliest window onto an Indo-Aryan-speaking society in India. In fact, we can call these people by the name they used of themselves: Aryans.

Aryan society had little in common with the urban civilization of Harappa. It was very much a society of pastoralists, proud of its cattle, its horses, its chariots, and its raids; it included kings and warriors alongside its priests, but had nothing we would really want to call a state. This society did not exist in a human vacuum, for alongside the Aryans there appears a dark-skinned people whom they hold in contempt. Thus it would not be hard to imagine this aggressive pastoral population playing a part in the demise of the Indus Valley civilization. What is surprising is the scale of its wider impact on India. It is not just that the language it brought to the subcontinent is now spoken in various forms by the majority of the Indian population. It is also that such rough-edged pastoralists should be at the root of Indian civilization as it has existed for the last three millennia. This is perhaps another instance of a lesson we learned from the ancient Near East: not to underestimate the potential historical impact of pastoral peoples.

Up to this point in our analysis the key role in the shaping of India has been played by the northwest. It was in the early centuries of the first millennium B.C. that this monopoly was finally broken. India was now entering its Iron Age, and iron tools made it easier to clear the jungle and plow the soil of the northeast; this in turn made possible the large-scale cultivation of an East Asian domesticate, rice (perhaps the only item of major importance to enter prehistoric India from the east). The outcome was a fundamental change in the human geography of the subcontinent: within a few centuries its center of gravity had shifted to the Gangetic plains. There the rise of cities toward the middle of the first millennium B.C. was matched by that of states. Our sources enable us to glimpse the existence of numerous states in competition, and in due course one of them, Magadha, came to dominate the entire region. A few of these states had been tribal republics (there were more such polities to the west, some of them sufficiently well developed to mint coins); but monarchy was dominant. At the same time the northeast was playing the key role in the formation of classical Indian civilization. This was the period in which India reacquired literacy. By the fourth century B.C., and perhaps for some time before, India was using a form of alphabetic writing that owed nothing to the forgotten script of the Indus Valley; this new script, Brāhmī, is the source of all the Indian scripts in use today.

A key development in the northeast around the fifth century B.C. was the emergence of Buddhism. Like other movements of the day that have left less of a mark on history, Buddhism was at heart a philosophy meant for ascetics who—like the Buddha himself—had renounced the world and wished only to be rid of it. For now we can leave aside the doctrinal content of this philosophy. What needs saying here is that the history of Buddhism (and not only Buddhism) highlights the fact that renouncing the world can be an effective technique for flourishing in it. In the first place, Buddhist ascetics were not, for the most part, loners. Instead, the founder left behind him a community of monks (to which he was persuaded, much against his better judgment, to add nuns). The community failed to hold together in the centuries after his death, despite a series of councils; but the various sects into which it split conformed to the same basic organizational pattern. Just as in Christianity, the existence of monks led to monasteries, and monasteries with some embarrassment became centers of wealth and power. What generated the wealth and power was the success with which the monks engaged the world they had renounced. They were adept at securing the patronage of rulers, who for a millennium found Buddhist monks no less eligible than Brahmins as providers of the religious endorsement without which it is hard for a king to look good. At the same time they ministered to the religious needs and wants of the laity, which was in no more hurry to renounce the world than its rulers were. In these respects the Buddhists were not so much different from their competitors (the Jainas, for example) as more successful; already at the second council, according to tradition, the main issue was whether monks could accept donations of gold and silver.

The Buddhists also did something their competitors did not do: they exported their religion beyond the frontiers of India. Anyone who renounces the world is free to disentangle himself from the parochial loyalties of the society he professes to abandon. Renouncers were thus under no obligation to respect ties of caste, ethnicity, or language. It was the Buddhists who made effective use of this freedom; to spread their religion, they first translated the massive literature they transmitted into several Indo-Aryan languages, and in due course they rendered it into such exotic tongues as Chinese, Tibetan, and Mongol. The Buddha, it was later said, could express everything he wished in any language whatever. At the same time the Buddhist missionaries had no quarrel with the native gods of the societies into which they moved. Hence, as we will see, Buddhism played a prominent role in the export of Indian civilization.

In one crucial respect, however, the emergence of the northeast has the air of a very conservative process. It perpetuated the Aryan tradition of the northwest, as is clear from the prevalence of Indo-Aryan languages in the region, and from the fact that Aryans dominated the caste system. It is not obvious that this had to be so. We could easily imagine some indigenous non-Aryan people combining iron and rice to develop the northeast in a quite different ethnic and cultural style. Had this happened, it is hard to say what future the culture of the Aryan pastoralists of the late Bronze Age would have had; it would probably have been as irretrievably lost as the overwhelming majority of the cultures of the human past. But as it was, the Aryan heritage, filtered through the urban civilization of the Gangetic plains, was now set to become the dominant cultural tradition of the entire subcontinent.

The south was on balance a significantly different story. If the northeast came onstage in the early first millennium B.C., it was only late in that millennium that the process got under way in the south. Here too the emergence of states was a crucial development, and states have uses for civilization. As might be expected, they adopted that of the north; like the Nubians in relation to ancient Egypt, the peoples of the south can scarcely have known any other. This meant large-scale importation of the appurtenances of north Indian culture—notably Brahmins, their Vedas, and their literate skills. There may also have been an element of military invasion from the north. But to judge by the outcome, the deployment of Aryan power cannot have been nearly as oppressive as in the northeast. At least in the areas we now think of as the south, the indigenous peoples retained their Dravidian languages and sooner or later used them to develop literary cultures of their own. They likewise preserved their non-Aryan ethnic identities, and despite the significant roles played by immigrant Brahmins, their political elites were predominantly non-Aryan. Only in Ceylon was a large territory permanently colonized by an Indo-Aryan-speaking people, the Sinhalese; and the key to this may be that they got to the island before the Dravidian-speakers.

This survey should not be taken to mean that the northwest had ceased to matter in Indian history. It remained the gateway to the subcontinent down to the eighteenth century A.D. Through this gateway came a succession of invaders, starting with the Persians in the sixth century B.C. and the Greeks in the fourth. Typically the political domination of these intruders was limited to the northwest, a pattern that still held good as late as the Muslim invasion of the early eighth century A.D. Culturally the impact of such invaders was usually absorbed or contained without serious dislocation. Thus the famous Sanskrit grammarian Pāṇini is likely to have lived under Persian rule. The Muslim presence was culturally less benign, and from the eleventh century onward Muslim conquest was to become a major threat to the traditional culture of India.

Classical Indian Civilization

Up to the end of the first millennium A.D., we can think of India as the domain of a single civilization. Not everybody in the subcontinent was part of this civilization. Down to the present day there are significant tribal populations that are outside Hindu society, and there were probably more in the past; some of them are hunters and gatherers. Moreover, many aspects of the civilization varied from one part of India to another even at an elite level. Regions differed, for example, as to whether months began at new moon (as in most of the south) or full moon (as in most of the north). At a popular level the regional variations must have been even greater. But there was still an overarching cultural unity. Brahmins, for instance, were quite often on the move; we know of Kashmiri Brahmins from the far north who settled among the Tamils in the far south. Manuscripts also traveled; Sanskrit texts written in the south in the script of Kerala show misreadings that could have arisen only in the process of copying from originals written in the script of Kashmir. By contrast, there was no parallel to this pan-Indian Aryan and Sanskrit role in the ethnic and linguistic makeup of the Mesoamerican cultural zone.

Cultural unity was not, however, matched by political unity. From time to time Indian history had its imperial episodes, as when empires were created and sustained by states based in the northeast. The Mauryas from the fourth to the second century B.C. were the earliest example of this, and the Guptas from the fourth to the sixth century A.D. were another. But these empires did not extend to the far south. The empire builders of the second millennium A.D. did better in this respect—but they were Muslim or British. This meant that some of India at all times, and most of it at most times, was divided among a plurality of regional states. Yet, as long as these states lavished patronage on the Brahmins, the overarching cultural unity of India was not in danger.

It is a frustrating, but also interesting, feature of Indian history that our knowledge of the fortunes of these states, large and small, is rather poor. For pre-Muslim India we have no equivalent to the rich historiography of China or the Islamic world. Instead, the history of these states must be reconstructed from the evidence of inscriptions, coins, and stray references in the literary sources. Usually this evidence is fragmentary at best, though there is one notable exception: the Mauryan emperor Aśoka in the mid-third century B.C. is a distinctly historical figure. We owe this to the number of surviving rock inscriptions in which this great patron of the Buddhists and their competitors explains himself in an idiom of vegetarian universalism. Thus he tells us that formerly hundreds of thousands of animals were slaughtered daily in the royal kitchens, but that this has now been reduced to two peacocks and sometimes a deer, and that even this will be phased out; he describes his remorse at the mass killing and deportation that had accompanied his conquest of the independent state of Kalinga; and he affirms that it is his duty to promote the welfare of the whole world, and so discharge his debt to all beings. But Aśoka is unique, and his inscriptions, for all their interest, are no substitute for a good chronicle. Why, then, do we have so little in the way of historical writing to inform us about the course of events in India? The obvious explanation would be that Indians did not write it. This would not be unique: the ancient Egyptians did not produce anything that could really be called a chronicle. But a Chinese visitor of the seventh century A.D. refers to Indian officials charged with the writing of history, and the problem may rather have been that Indian society, and more especially its Brahmin cultural elite, did not preserve what these historians wrote once the dynasty they served had come to an end. Given all that the dynasties had done for the Brahmins, it seems curiously ungrateful.

What this society did preserve, and in enormous quantities, was a literature of a broadly religious character. Since this preservation turned on continuous transmission, the texts available to us today are by and large those of the schools of thought that survived. But the results are nothing to complain of. The tradition of the Brahmins, a key component of mainstream Hindu religion, preserves not just the four Vedas but a great variety of later Sanskrit texts of diverse content; ancient Indian atomism, for example, is well represented (there is nothing specifically modern about the idea that matter is made up of atoms). Buddhism eventually disappeared in India itself, but a large Buddhist literature originally composed there still survives among Buddhists in Ceylon and elsewhere in a variety of languages. Jainism, a religious movement of the same character as Buddhism but slightly older, is today confined to two widely separated communities in India; one of them has preserved a literary heritage reaching back into the first millennium B.C. The documentation for the religious history of ancient India is thus as rich as that for its political history is poor. What this material suffers from, as might be expected, is a certain historical disembodiment. For example, the Buddhist scriptures of Ceylon are in an Indo-Aryan language known as Pāli; it stands to reason that this must once have been the vernacular language of real people living in a real place, but we can only guess where in northern India that might have been.

The survival of Indian Buddhist texts outside India brings us to the final theme of this survey: the spread of Indian civilization to regions beyond the subcontinent. This did not involve conquest. The northwest, in fact, has always been a gateway through which foreigners invade India, and never the other way around. Instead, the process seems to have been rather similar to the reception of north Indian culture in the south. We see it at work in two very different areas in the early centuries of our era: the Asian interior and Southeast Asia.

The enduring legacy of Indian influence in the Asian interior was the spread of Buddhism to China, and thence to other parts of East Asia. China, however, already had a civilization of its own, and so was in no need of borrowing India's. But it was just such a process of borrowing that had brought Buddhism from northwest India around the top of Tibet to the borders of China. In this region we have evidence from the second century A.D. of culturally Indianized states among peoples that had previously lacked literate culture. Several centuries later something similar was to happen in Tibet.

In roughly the same period, a century or two after the turn of our era, Indian civilization began to make its influence felt in parts of Southeast Asia. The precondition for this was the rapid development of long-distance trade on the Indian Ocean, dramatically illustrated by the presence of Roman artifacts at Go Oc Eo in what is now Vietnam (including a gold medallion of the emperor Antoninus Pius dating from A.D. 152). As in southern India, the process meant a dramatic increase in employment opportunities for Brahmins of fortune. A fifth-century Chinese source mentions the presence in one Southeast Asian kingdom of over a thousand Indian Brahmins, to whom the local people would give their daughters in marriage. But there seems to have been less Indianization of society at large; thus, unlike the peoples of the Indian south, those of Southeast Asia do not have well-developed caste systems. So it was primarily the elites and rulers of the region who had a use for Indian civilization. This use was nevertheless considerable. Inscriptions attest the emergence of Indianized states from the third century A.D., and by the end of the millennium numerous such states had appeared on the mainland and some of the islands.

We can end this survey by noting an interesting asymmetry about the export of Indian culture: it achieved its successes to the east, and not to the west. To the east, peoples whose geographical locations gave them a choice between Indian and Chinese culture overwhelmingly chose Indian. The Tibetans provide a further example of this preference: we know that they were at one time interested in both cultures, but they later dropped the Chinese option.

By contrast, there was no such spread of Indian culture to the west. The Near East, of course, already had literate culture, just as China did; but the massive spread of Buddhism in China has no parallel in the Near East, even though Mani respectfully included the Buddha in his synthesis.

II. The Nambudiri Brahmins of Kerala

Kerala is the rainy coastal strip of southwestern India. To the east lies the Tamil country, of which it was formerly a part; Malayalam, the Dravidian language of Kerala, began its life as a Tamil dialect. The society of Kerala is a highly differentiated patchwork of castes and subcastes, so much so that the region has been described as a "madhouse of caste." One small, intriguing, and formerly very prestigious element in this patchwork was the Nambudiris; they numbered about sixty thousand in the middle of the last century. Taking a closer look at them is one way into the labyrinth of caste.

The Nambudiris are Brahmins and, as such, the distant heirs of the Aryan ritual tradition of the Bronze Age Punjab. Those Nambudiris who maintained their Vedic traditions did so much more faithfully than most Indian Brahmin communities of modern times. They had an educated knowledge of Sanskrit; between them they transmitted three of the four Vedas, together with a large amount of associated cultic material; and they performed complex Vedic rituals. All this took a great deal of training. A Nambudiri boy embarked on the memorization of his family's Veda, whichever it was, at a tender age. The boys later learned to recite their texts forward and backward in elaborate patterns; the effect was to ensure the continuing oral transmission of an unchanging text. This fidelity was not in vain. Comparative philologists have inferred the existence in the Indo-European mother tongue of a raised accent that has disappeared in all surviving daughter languages; the only place where it lived on was in a special form of Nambudiri Vedic recitation. Less dramatic, but of interest to Indologists, is the fact that the Nambudiris transmitted a recension of the Ṛgveda that was preserved nowhere else.

Yet the Nambudiris of the last century were not just a living fossil. They were distinctive among Indian Brahmins not only for what they preserved but also for what they had become. They had customs that diverged from those generally accepted by Brahmins elsewhere in India (one of them, as we will see, fundamental to their relationship with the indigenous society of Kerala). The standard view of such customs was that in principle there could be only one correct way for Brahmins to do things all over India; but in practice regional divergences existed and had to be accepted, as with the meat-eating Brahmins of Kashmir. The Nambudiris were said to have sixty-four such practices (note, incidentally, the multiple of sixteen). Moreover, in some respects they had assimilated quite drastically to the Keralan milieu. Malayalam was their mother tongue, and it deeply influenced their pronunciation of Sanskrit (far more so than Tamil did among the Brahmins farther to the east). In many such ways they belonged far more to contemporary Kerala than to the pastoral Punjab of the Ṛgveda.

How did the Nambudiris succeed in becoming an organic part of Keralan society without losing their Aryan heritage? It is always possible to combine two cultures; the problem is to ensure the long-term stability of such a combination. Here history shortchanges us. We do not know when the Nambudiris arrived in Kerala, though they were there by the ninth century A.D., or where they came from, though directly or indirectly it was doubtless from somewhere to the north; and we are not sufficiently informed about their early life in Kerala, though it is clear that royal patronage, and the agricultural wealth it placed in the hands of the Nambudiris, was crucial to their success. But traditional Nambudiri society as known to us in modern times had solved the problem of cultural stability in an ingenious manner, and the elements of the solution were certainly old. One of the peculiar customs of the Nambudiris was that only the eldest son was allowed to contract a legal marriage; this meant that only he could marry a Nambudiri woman, so only his children could be Nambudiris like their father. This was crucial, since only Nambudiris would inherit the family land. The effect was to ensure that the transmitters of the Vedic tradition were an endogamous lineage guaranteed the wealth they needed to lead a life of cultured leisure. Or at least this was the effect until the system was undone in modern times.

This solution left two groups at an obvious disadvantage. One was the younger sons. But fortunately the system provided for them in a slightly irregular way. The Nambudiris enjoyed good relations with a respected indigenous military caste, the Nayars, who were prepared to give them access to their daughters; and unlike other Brahmin groups, the Nambudiris were willing to countenance this. The resulting matches were properly a form of concubinage, not marriage. Children born to the couple were not recognized as Nambudiris; instead, they were Nayars, belonging to the same matrilineal group as their mother. They would not, of course, inherit Nambudiri land, nor would they participate in Nambudiri rituals (it is not for a Nayar to recite, or even hear, the Vedas). From the point of view of the younger son, this arrangement had its inconveniences: strictly speaking, the rules of ritual purity prevented him from eating food cooked by his Nayar consort, or from eating or bathing with his Nayar sons. But an arrangement of this kind was a lot better than nothing. In fact, in traditional Kerala it was a way to develop valuable connections with prominent Nayar families, and it was in considerable measure through such arrangements that the Nambudiris cemented their position in Keralan society. As one Nambudiri summed up his own family's relationship with a Nayar family in the 1960s, "The women were pretty, they fed the husbands well [!], and they were useful people to be close to."

The other group adversely affected consisted of the Nambudiri girls, who faced a dearth of potential husbands. Here, of course, no analogous arrangement could be made, since it would have been out of the question for the Nambudiris to give their daughters to men of a lower caste. This did not, however, mean that surplus daughters were condemned to spinsterhood: a Nambudiri eldest son might have up to three wives at a time, and a girl could always be married off to an old man one of whose wives had died. But one senses that the lot of a Nayar concubine may have been happier than that of a Nambudiri bride. The Nayar concubine continued to live with her own family, and to be very much a part of it, whereas the ties between a Nambudiri woman and her original family tended to be severed at her marriage, leaving her isolated among her in-laws. Moreover, the Nayar concubine, unlike the Nambudiri wife, could have several consorts at once, and within certain limits she had considerable freedom in choosing them; but this practice, as might be expected, went out of fashion in Victorian times.

We have here one small example of the workings of a caste society. Some of its themes are specific to the relationship between the Nambudiris and the Nayars, but others are part of the bundle of features that tend to appear together in any account of an Indian caste system. Society is divided into named castes, apart from groups like mountain or forest tribes that are outside the system altogether. The system, while encouraging a remarkable degree of social diversity, tends to be strongly hierarchic: as we have seen, the Nambudiris were a higher caste than the Nayars, who in turn outranked many other castes in Kerala. Membership is determined by birth: you cannot choose to join a caste, or be elected to one (though you may be expelled from one). Marriage is constrained by caste, with many castes being endogamous. Occupation is linked to caste, albeit often loosely: Nambudiris are priests, Nayars are warriors (there are even criminal castes in India, just as there are criminal gods). Conceptions of purity and pollution play a key role in keeping castes apart: as we saw, a Nambudiri is not supposed to eat food cooked by a Nayar, and such a rule commonly obtains between higher and lower castes. In fact, each of these features could be paralleled many times over in the Indian context.

Alongside these recurrent features must be set a fact of a different order: the underpinning of the system by religion. According to an ancient tradition going back to the Vedas, society was divided into four classes (literally "colors"). At the top was the priestly class (the Brahmins), then came the warrior class (the Kṣatriyas), and then the economically productive class (the Vaiśyas); all these were Aryan. At the bottom was the non-Aryan servant class (the Śūdras). This schema may once have been a more or less straightforward description of a particular society, but this was not why it mattered for most of Indian history. What it provided was rather an archetype in terms of which Indians at different times and places could understand, criticize, or justify the particular caste systems they lived in. For example, the multiplicity of actual castes was explained as the result of a variety of irregular crossings between the four original classes. It was not just Brahmins learned in the Vedic tradition who made use of the schema; Kabīr Das, a late medieval poet in a popular religious tradition who had not the slightest use for the Vedas, articulated his message by singing of a land in which there is no Brahmin, no Kṣatriya, no Vaiśya, and no Śūdra.

How uniquely Indian is caste? Bīrūnī, an eleventh-century Muslim observer, contrasted the Indian caste system with Muslim religious egalitarianism, which recognizes interpersonal inequalities only in piety. He remarked that this difference was the greatest barrier interposing between the Indians and Islam. It is easy for someone socialized into the secular egalitarianism of the modern West to react in a similar way. But if we go back to the bundle of features typically associated with the system, it is evident that none of the underlying ideas is peculiar to India. We are used to societies containing a variety of groups that vary in prestige. Such groups may be linked to birth (you can hardly choose to become a Gypsy if you are not one already) and to occupation (a Boston Brahmin does not make a living by cleaning toilets). They also make a difference to marriage (you are more likely to encourage your daughter to marry a member of a high-ranking than a low-ranking group). Though it is impolite, it is not unknown for people in many societies to speak of other groups as dirty. So at this level we could hardly claim that the elements of the Indian caste system are beyond our understanding.

There are, moreover, some significant, though partial, parallels to the workings of the system in other societies. These tend to be most visible at the extremes, as with aristocracies of blood at the top and outcaste groups at the bottom; it is no accident that the two Indian caste terms that have entered standard English vocabulary are "Brahmin" and "Pariah" (the Paraiyan being a Tamil outcaste group that traditionally played drums at funerals to keep evil spirits away). Here let us focus on the outcastes. In terms of the orthodox schema they ranked below the Śūdras. Members of one such category in ancient India were not allowed to live in the same settlements as Aryans, and had to eat from broken vessels; they were so polluting that they came to be required to warn Aryans of their approach by striking a wooden clapper. In more recent times outcastes have represented some 15 percent of the Indian population and, like everyone else, have been divided into numerous castes of their own. Thus in Kerala the Pulayas, who traditionally could be bought and sold, were obliged to call out every few paces when walking along a path to warn others of their polluting presence; if they heard an answer in a higher-caste voice, they had to step down into the ditch. But at least they were better than the Nayadis—even to see a Nayadi was polluting to members of superior castes. Now, if we leave aside such details, it is not hard to find groups in other societies to which we could readily apply the term "outcaste" in the same spirit, in order to convey the exclusion and humiliation to which they are subjected. A famous East Asian example is the group traditionally known in Japan as the Eta; a similar population in Korea has attracted less attention. In Europe the Gypsies are as good an example as any.

What, then, makes India different? The answer seems to be two things. The first is that the familiar elements have been put together into a formal system that shapes the society from top to bottom. In Japan, by contrast, it is taking a certain liberty to call the Eta "outcastes," since there is no broader set of castes for them to be out of; and in general the non-Indian parallels tend to leave us with the sense that they are fragmentary and incomplete. The second thing that distinguishes India is the depth of the traditional religious underpinning of the caste system. What, then, is the relationship between these two things? It is certainly possible for them to exist apart, as is shown by the contrasting cases of two islands. One is the neighboring island of Ceylon or, more precisely, the Sinhalese highlands of the interior: here caste is alive and well as a social institution in a population of Indian origin, despite more than two thousand years of Buddhist doctrinal indifference to it. The other island is Bali in Southeast Asia, with a population of a quite different origin: despite its long-established Hinduism, it has only the semblance of a caste system. But over the mass of the Indian subcontinent, the association of caste as a social system with the Hindu religious tradition is ancient and intimate; and though the two can now live apart, they give the impression of having evolved together.

The obvious question that remains is why the social structure of India should have developed in this distinctive way. To this, as to so many such questions, we do not have the beginnings of an answer. The one thing we can say is that the rather isolating geography of the subcontinent lends itself to the evolution, spread, and preservation of cultural idiosyncrasies not shared by the rest of Eurasia. So we have no reason to be surprised that Indian society should have gone off in a direction of its own. But why that direction rather than some other?

III. Of Gods and Courtesans

At Konārak in Orissa, a large and imposing temple of the sun god was built in the thirteenth century A.D., and has since come to be known as the Black Pagoda. This temple is famous, or notorious, for the unabashed sexuality of its sculptures. In itself, erotic sculpture is nothing unusual in the decoration of an Indian temple; couples are freely represented in more or less intimate embrace. But by common consent the Black Pagoda takes the prize. In its days of glory, however, its sculptures may have fought a losing battle for the attention of male worshipers. Medieval Indian gods kept court in the same style as medieval Indian kings, which meant that they had to be attended by beautiful and accomplished courtesans. These courtesans would sing and dance before the god, and might also be willing to bestow their favors on his worshipers. So when a general in the south founded a temple in memory of his mother around A.D. 1100, he equipped it with quarters for the most exquisite courtesans. Likewise, the temples of medieval Kerala, which were bastions of Nambudiri wealth and power, were well provided in this respect. Perhaps, then, the sculptures at the Black Pagoda were an advertisement for the temple's courtesans. Or perhaps they were edifying symbols of mystical union with the divine. Or perhaps they were both. The fact is that we cannot tell, since those who installed the sculptures have left us no record of their intentions. Once again we lack a soundtrack.

Before we go looking for one, we should return for a minute to the long and ramified history of Indian religion. As a first approximation it helps to think of this in terms of three major developments, two of which are already somewhat familiar.

The first was the Vedic tradition transmitted by the Brahmins. This, as we have seen, went all the way back to the second millennium B.C., and it remains alive today. Its keynote was ritual. Those who like such things can derive tremendous satisfaction from faithfully reenacting every detail of an ancient cult, even one whose original meaning may have been forgotten long ago. To others a tradition of this kind may seem intellectually meaningless and spiritually empty.

The second development was the philosophical style of religion that emerged in the first millennium B.C. on the Gangetic plains. This new way of thinking arose within the broad mainstream tradition transmitted by the Brahmins, but it also took the form of new religious movements, notably Jainism and Buddhism, that cut free from the Vedic heritage. Initially at least, such religions tended to offer their adherents materialist accounts of the world in which the role of the gods was somewhat trivial. The bad news they conveyed was that life is an endless cycle of rebirths; the good news was that if you renounced the world and practiced a suitable asceticism, you might eventually be able to end your participation in this tedious cycle and achieve extinction. This way of thinking provides a systematic cosmic vision on a grand scale, something missing from the original Vedic heritage, and for anyone who yearns to be extinct it is purpose-built. But again, not everybody feels this way; most of us would readily settle for a better deal in the next life, be it as a film star or an Olympic athlete.

What both these forms of religiosity tended to lack was emotional warmth. This was to be the central contribution of the third major development in the history of Indian religion, the rise of the Hindu devotionalist cults in the first millennium A.D. In this process the south played a leading role, though its roots went back to developments in the north in the preceding millennium (they include a religious poem that in modern times has become famous even outside India, the *Bhagavad Gītā*). The adherents of such a cult loved their god, so much so that often they would worship no others; and their god loved them in return, so much so that he would forgive their sins and even become incarnate for their sakes. This, then, is another case of monolatry, and it could accordingly develop a virulent strain of intolerance. One thirteenth-century devotee of Śiva refused even to set eyes on those who were not members of his sect, let alone touch or talk to them; he held that books referring to Śiva in derogatory terms should be burned without hesitation and their authors killed. Both the devotion and the intolerance are familiar from Christianity; perhaps the main difference is that whereas a Hindu god may endure incarnation repeatedly out of love for his followers, for the Christian god once was enough.

Not surprisingly, profane love provided a ready model for imagining divine love. But profane love comes in various shapes and sizes; which, then, is the appropriate model? Devoted Christians, for example, adore Jesus, but they do not flirt with him, let alone aspire to sleep with him. Some early Tamil devotional poetry would fit quite well with a Christian (though not a feminist) sensibility: the worshiper is represented as a woman who lives forgotten and loves forlorn, pining for her absent lover. The profane world invoked by the poet is thus one in which "those born as women see much grief," and there is little they can do but sigh and endure it.

In a later devotional tradition, however, we encounter a quite different sensibility. The poems in question were composed in Telugu, another Dravidian language of the south, in the fifteenth to the eighteenth century. Two rather extreme examples will make the point. In one poem, with the refrain "Handsome, aren't you?," the woman tells the god that he may be the prince of playboys, but that he is still not going to get her on credit: "You can make love like nobody else, but just don't make promises you can't keep. Pay up, it's wrong to break your word." In another, she tells him that he may enter her house, but only if he has the money. She then sets out her tariff; "total union" is the most expensive item, and for this she requires to be bathed in a shower of gold. Here the profane reference is not to the tedium and constraint of respectable womanhood but to the sexual freedom of the courtesan. This is no accident, for it was in the first instance the courtesans in the temples who sang these Telugu poems. Later the genre was taken up by the courtesans at the royal court, an easy slippage since the kings of the day were considered divine. In either setting the poems could be taken on more than one level. Perhaps there was something a bit similar at Konārak.

How exotic is all this by European standards? As we already noted, we need look no further than Christianity for the analogy between sacred and profane love. What is different about the Telugu poems is that they are very sexy—as only profane poetry is expected to be in a Christian society. This in turn is related to the institutional setting. Courtesans as such are, of course, in no way an Indian monopoly. But temple courtesans, with their dual role as companions of gods and men, merge the sacred and the profane in a way that is more unusual. Not that such arrangements were unique to India: temples provided comparable amenities in the ancient Near East. There were male temple prostitutes (doubtless for men, not women) in the temple in Jerusalem until King Josiah destroyed their houses in the later seventh century B.C. (2 Kings 23:7). In the same vein the Bible prohibits the profession of temple prostitute to Israelites, male and female alike (Deuteronomy 23:17); yet the very terms it uses for such prostitutes imply a sacred status. Even in Christianity and Islam, going on pilgrimage remained a good strategy for a man interested in spending time away from his wife. But there is no parallel in the later monotheist world to the full-blooded confluence of the sacred and the profane around a medieval Indian temple.

CHAPTER 9

CHINA

I. The Making of China

Like many parts of the world, China is geologically an assemblage of bits and pieces. North China is one block, and an appropriately ancient one—some of the oldest rocks in the world are to be found there. South China—or, more precisely, central and southern China—forms another block, itself the product of a merger. But these elements had become part of what is now Eurasia well before the breakup of Pangea into Laurasia and Gondwana, so that in this sense China, unlike the southern Near East or India, is an original part of Laurasia. It has nevertheless been strongly affected by the arrival of India through the uplifting of the Tibetan plateau, which like India it adjoins. The fact that both regions back onto Tibet means that they have a number of features in common; but as we can see by comparing them, the differences are just as important, and in one respect go back to the heterogeneous origins of China.

A first comparison concerns boundaries. India is pretty well delimited by the combination of mountains to the north and ocean to the south. In the case of China mountains do indeed mark the limit of the region to the west, and ocean does the same to the southeast and east (though the oceanic frontier is much more sensitive to changes in sea level than in the Indian case, with the result that China is much larger in an ice age). But this still leaves China with two significant land frontiers that are not blocked by high mountains. One of these frontiers is in the southwest, where China adjoins Southeast Asia; this is the shorter of the two, and movement across it is impeded by jungle. The other is the long northern frontier, which runs through open country. It is this frontier that has mattered most to Chinese history in terms of both threat and opportunity—the threat of conquest by nomads from the steppes to the north, and the opportunity for contact with other civilizations to the west.

Another comparison concerns the distribution of mountains and plains. Like India, China has its highlands, though again they are unimpressive in comparison to those of Tibet. But whereas India has a single block of highlands in the south and a single concentration of alluvial plains in the north, the makeup of China is more complex: as one moves from north to south, river valleys alternate with bands of highland terrain. Much as in India, it is two rivers rising from the Tibetan plateau that dominate the picture; but here both flow from west to east. North China is the land of the Yellow River. In the northwest it flows through massive windblown deposits of a yellowish soil known as loess (hence the name of the river). Farther east it picks an unstable course through a vast accumulation of silt that it has itself transported from the loess deposits upstream. Central China is dominated by the Yangtze, which has likewise created an extended alluvial plain. Both rivers are dangerous, but their valleys—especially that of the Yangtze—have remarkable agricultural potential. The highlands that separate the two are a residue of the joining of the two main blocks out of which China was formed. South of the Yangtze, China is not lacking in rivers with agriculturally exploitable valleys, but none are on the same scale.

A final comparison concerns climate. Like India, China is very much a part of monsoon Asia. It shares with India its overall tendency to dry winters and wet summers, the summer rain being brought by winds blowing off the ocean from the southeast. The geographical distribution of the rainfall, however, is somewhat different (see map 7). China resembles India in having an arid northwest, but the Chinese northeast is also relatively dry, so that the overall contrast is between a dry north and a wet south (though the north was less dry in the early Holocene than it is now). The most dramatic difference between India and China, however, concerns the severity of winter. Whereas Tibet protects India from the biting winds of Siberia, China is wide open to them. Winter in the interior is thus brutal in the north and cold even in central China.

The archaeological evidence suggests that it was primarily from the north that modern humans entered China. People with a culture of an Upper Palaeolithic character are attested in Siberia from about forty thousand years ago, and seem to have expanded southward into northern China, Korea, and Japan. South China, like Southeast Asia, is more or less a blank until the Holocene. But there was apparently a population of cave-dwelling Negritos on Taiwan down to the nineteenth century, and this could be a residue of an early coastal entry of modern humans from Southeast Asia. A few Negrito groups survive there, and bear more resemblance to the native peoples of New Guinea than to Southeast Asians of the present day.

It is with the Neolithic that China comes into sharp focus. Two major developments marked its onset. In the valley of the Yellow River, settled farming based on domesticated millet had appeared by about 6000 B.C. In the Yangtze Valley, domesticated rice was being cultivated by about 5500 B.C., and perhaps as early as 7500 B.C. (this seems to be the source of later rice cultivation in south-

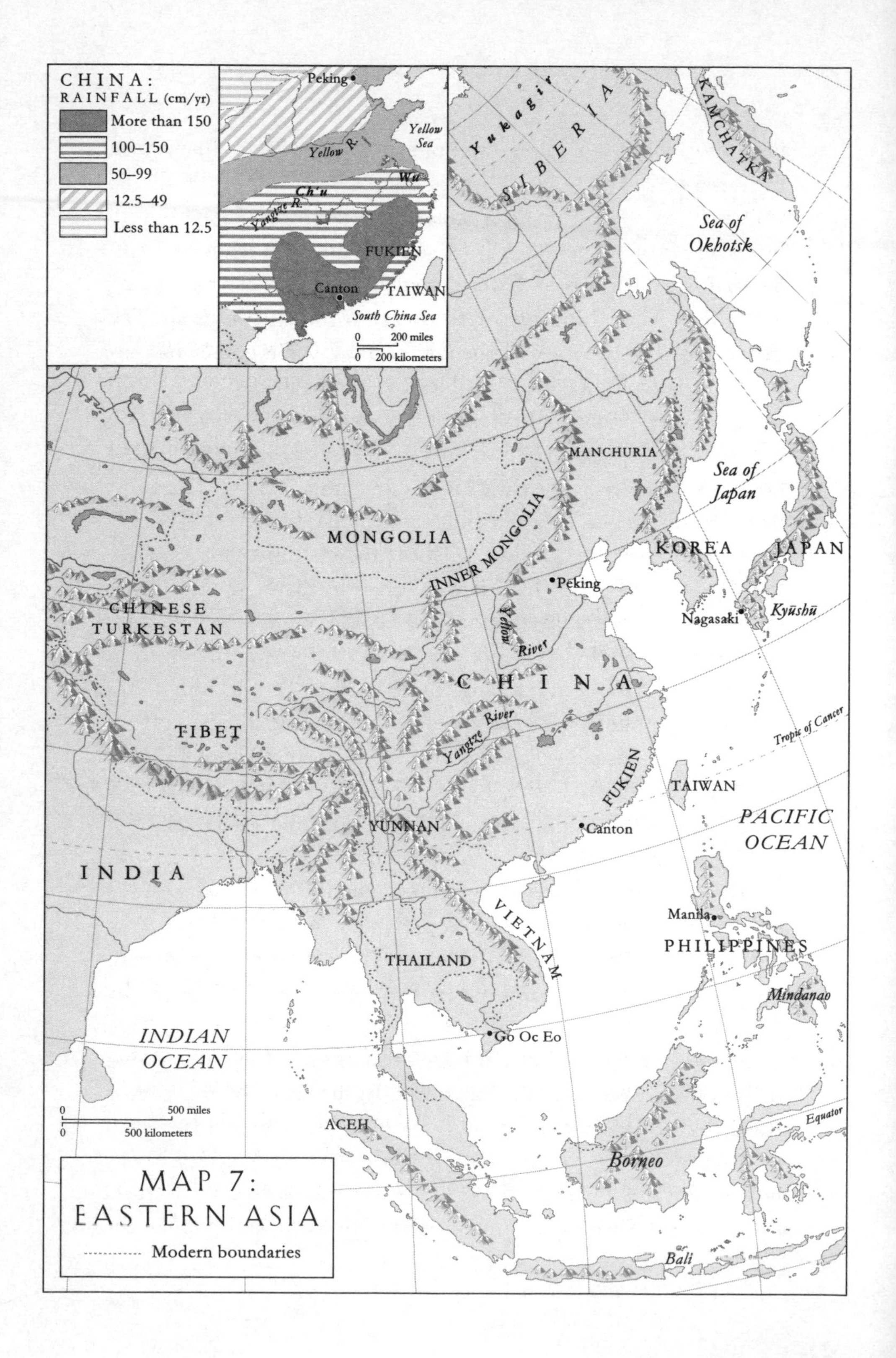

CHINA: RAINFALL (cm/yr)
More than 150
100–150
50–99
12.5–49
Less than 12.5
Peking
Yellow Sea
Yellow R.
Wu
Ch'u
Yangtze R.
FUKIEN
Canton
TAIWAN
South China Sea
0 200 miles
0 200 kilometers
Yukagir
SIBERIA
KAMCHATKA
Sea of Okhotsk
MANCHURIA
Sea of Japan
MONGOLIA
INNER MONGOLIA
KOREA
JAPAN
Peking
Nagasaki
Kyūshū
CHINESE TURKESTAN
Yellow River
CHINA
TIBET
Yangtze River
Tropic of Cancer
FUKIEN
TAIWAN
YUNNAN
Canton
PACIFIC OCEAN
INDIA
VIETNAM
Manila
PHILIPPINES
THAILAND
Mindanao
INDIAN OCEAN
Go Oc Eo
0 500 miles
0 500 kilometers
ACEH
Equator
Borneo
MAP 7: EASTERN ASIA
Modern boundaries
Bali

ern China, Southeast Asia, and India, though there is no certainty yet regarding where rice was first domesticated). As might be expected, the cultivation of these crops came to be associated with domesticated animals, notably the chicken, the pig, and the water buffalo; but it is not yet clear just when they were domesticated. The two crops differed significantly: the millet grown in the north was adapted to arid conditions, whereas the rice grown in the Yangtze Valley flourished in shallow water. But the overall pattern seems familiar enough from the onset of farming elsewhere. Yet one feature of the Chinese Neolithic is distinctly unusual: right from the start the major developments took place in the river valleys. This is not how it was in Egypt, Mesopotamia, or—so far as we can tell—the Indus Valley. Here perhaps is one reason why the subsequent buildup of settled village life was so strong in Neolithic China. By the end of the third millennium, solid agricultural foundations had been laid for the emergence of a civilization.

The Emergence of Chinese Civilization

Chinese civilization must have taken shape in the first half of the second millennium B.C. It was a product of the Yellow River valley, not of the Yangtze; rather as in India, the earliest appearance of civilization was in the part of China most likely to be in distant contact with the Near East and most similar to it. Though we are not well informed about the early centuries of this culture, we do have a reasonably good picture of what it looked like toward the end of the millennium, during the last phase of the rule of the Shang, a dynasty that must have come to power around 1600 B.C.

At this late stage of its history (about 1200–1050 B.C.) Shang culture combined elements of strikingly diverse origin. As might be expected, there was substantial continuity with the northern Chinese Neolithic. The use of stamped earth as a foundation for buildings is one obvious example, since it is well attested for earlier Neolithic cultures in the region. Another is the three-legged

Fig. 16: A *chia* of the early Shang period (*left*), with sketches of several types of bronze vessel made in Shang times (*right*; the one at the bottom is a *ting*).

design of some types of Shang bronze vessel (like several of those in figure 16); this is already found in Neolithic pottery.

At least one feature of the culture had its origins far to the west. This was the chariot, an instrument of warfare that had proliferated alarmingly in the early centuries of the second millennium. Shang chariots were in fact closest in the details of their design to those of Transcaucasia. We do not know the route by which the chariot reached China, but this transfer of military technology is quite likely to have been the work of Indo-European-speaking nomads; speakers of an archaic Indo-European language were still living to the northwest of China in the later first millennium A.D., and burials excavated in this region suggest that their ancestors were already there in the second millennium B.C.

There were also crucial elements in the late Shang complex whose origins are harder to determine. One is bronze. It is quite conceivable that Chinese bronze working was an indigenous

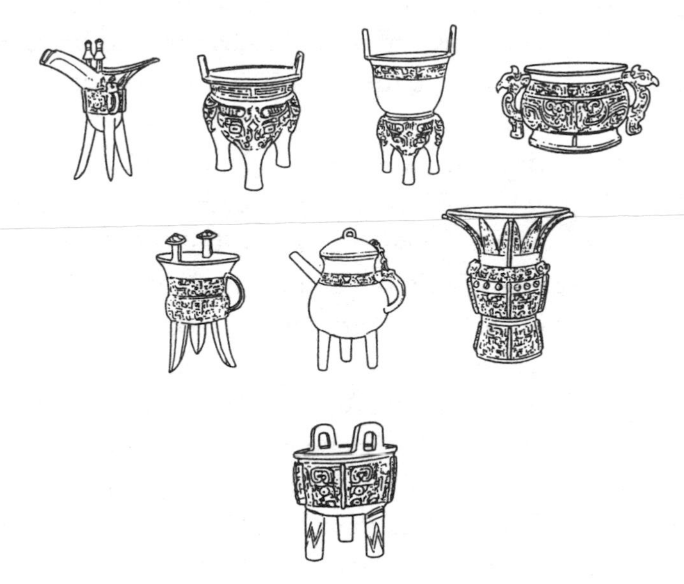

development, since Neolithic China already possessed the technology for firing pottery at high temperatures. But it could also be an import from the west. It is suggestive that the earliest evidence of bronze in China comes from the northwest, where it dates from about 2000 B.C., a millennium after the development of bronze in the Near East and half a millennium before its appearance in the Yellow River valley. Another such element is writing, which survives on large numbers of inscribed oracle bones and in some short inscriptions found on bronzes. Did the Chinese script emerge independently, as was clearly the case with Mesoamerica? Or was it influenced by the existence of writing elsewhere, as may have been the case with Egypt and the Indus Valley? In any event, the new technology was in each instance combined with long-established Neolithic traditions; for example, oracle bones had been used in divination long before the Shang began to write on them.

If we look back from the late Shang period, it is frustrating not to know when and how these and other features of Shang culture coalesced. Did the Shang always have chariots and writing, for example, or did they acquire them only at some secondary stage in their history?

But if we look forward rather than back, the view is unimpeded. Unlike the Near East and India, China never had occasion to rub out its earliest civilization and start again. Instead, we confront a remarkable case of cultural continuity: this civilization is the only one in the world that has existed continuously since the second millennium B.C. This does not mean that the Chinese of a later millennium would have felt particularly at home in the world of the Shang; cultural continuity, like memory, is selective. But the script of China at the beginning of the third millennium A.D. is a descendant of that used in the Shang capital at the end of the second millennium B.C., and the same relationship holds with regard to the language. This continuity is crucial to such ability as we have to make sense of the written remains of late Shang culture. It means that characters and the words they represent can often be identified with those of later and better-known periods. It also means that the later literary sources preserve at the very least a skeleton of Shang history that can be put to work in interpreting the Shang texts. For much as in ancient Egypt and Mesopotamia, one use of literary continuity was to preserve a historical memory centered on the succession of dynasties that ruled the country—except that by modern times the length of that memory in China substantially exceeded what either the Egyptians or the Mesopotamians had achieved.

The dynastic backbone of subsequent Chinese history can be conveyed in bold outline as a series of alternating periods of unity and disunity. The first period of unity was under the early Chou (or Zhou, from around 1050 to 771 B.C.); then followed a long period of disunity (771–221 B.C.), during most of which the nominal rule of the Chou lingered on. The second period of unity was initiated by the Ch'in (or Qin, 221–206 B.C.) and continued by the Han (206

B.C.–A.D. 220); then came several centuries of unadorned disunity (220–581). The third period of unity was started by the Sui (581–618) and maintained by the T'ang (or Tang, 618–907); then followed a few decades of disunity (907–60). The fourth period of unity (960–1127) was the work of the Sung (or Song), but the later part of their rule (1127–1279) was again a period of disunity. The final period of unity lasted through to the end of the history of traditional China, under the Yüan (or Yuan, 1260–1368), the Ming (1368–1644), and the Ch'ing (or Qing, 1644–1912).

This schema, crude as it is, alerts us to two interesting features of Chinese history. The first is that in China, in marked contrast to India, political unity is the norm; we can subsume much of the Chinese past under the rubric of "imperial China," whereas we cannot do the same for India. The second feature is that over the millennia the periods of disunity tended to become shorter; China became more, not less, imperial. But the schema, however useful as a first approximation, also conceals as much as it reveals, even at the level of political history. The single most important thing it fails to register is the watershed in Chinese history represented by the Ch'in unification of the third century B.C.

The Ch'in Watershed

The significance of this watershed is perhaps most obvious in terms of the political geography of China. The China of Shang times may have been no more than a part of the Yellow River valley; its neighbors could well have been non-Chinese. The Chou political order covered a much larger region of the north, but did not extend to the Yangtze Valley to the south. In this sense China down to the third century B.C. was still what we now think of as the north. By now, however, the people of the Yangtze Valley in central China had been strongly affected by Chinese civilization, either because it was brought to them by invaders from the north or because their own elites were rapidly assimilating it. In the middle of the first millennium B.C., the state of Ch'u (or Chu) in the middle Yangtze Valley looks like an example of the first process, whereas the state of Wu in the lower Yangtze region looks more like the second. But these processes may not have sufficed to make these regions fully Chinese. Even in the late third century B.C., there was a view that the men of Ch'u were "nothing but monkeys with hats on"—though one man who was injudicious enough to remark on this was boiled alive for his pains. And beyond the Yangtze there were vast non-Chinese territories to the south. The significance of the imperial paradigm established by the Ch'in was that it had the effect of placing the southward expansion of Chinese culture on a quite different footing, combining direct northern domination with immigration on a large scale. It still took time; as late as the ninth century A.D. for a northern bureaucrat to be banished to the far south was almost a fate worse than death. But in the last millennium Canton has been as much part of China as the Yellow River valley.

The contemporary language map of China brings out the degree of this incorporation. Over most of central and southern China, non-Chinese languages have been reduced to small islands in a sea of Chinese; only in the far southwest have indigenous languages held out to a significant extent. At the same time the distribution of the Chinese dialects—we could really call them languages—has something to tell us. Mandarin is spoken over a vast area in the north and west, whereas the Chinese of a wide region in the southeast is fragmented into very different dialects. In a way this is paradoxical: since Chinese comes from the north, that is where, other things being equal, it ought to be most deeply differentiated. Clearly other things have not been equal. One aspect of this is environmental: the open plains of the north are a homogenizing environment, in contrast to the fragmenting landscape of the south. The other major point is that it has typically been the north rather than the south that has constituted the core of the Chinese state; and Mandarin, as its name suggests, is the talk of officials. But these facts make it all the more remarkable that the southeast should have come to speak Chinese at all. Moreover, the study of the dialects spoken in this region can give us a sense of when the process occurred: with the exception of the dialects of Fukien (the coastal region opposite Taiwan), they seem to descend from the standard Chinese of the T'ang dynasty.

All this is in stark contrast to what we find in India. A central theme in the formation of the Indian culture zone was the adoption of northern civilization by the politically independent, ethnically distinct states of the south. In the end such a process was to play only a limited part in the making of China: Wu, if it was a case of native elites importing Chinese civilization, turned out to be the exception rather than the rule. It was only in a few areas beyond the reach of sustained imperial rule, or of any imperial rule at all, that such a process had lasting results. Northern domination made the rest ethnically Chinese. The difference is writ large on the linguistic maps of India and China. Whereas non-Chinese languages are residual in southern China, non-Indo-Aryan languages dominate the south of India. Likewise the breakup of Chinese into regional languages in the south finds its Indian parallel in the linguistic fragmentation of the Indo-Aryan north, and by the same token traditional India had no language comparable in spread to Mandarin.

If the Ch'in watershed was fundamental to the political geography of China, it was also of central importance for its cultural history. One aspect of this arose from Ch'in attitudes to the past. All ancient cultures are subject to gradual attrition as bits of their heritages go missing—a ritual here, a book there. (In China before the invention of paper in Han times, books were written on strips of bamboo held in order with string; imagine what happened when the string broke.) But the Ch'in experience was more like a bottleneck. It was not just military and political upheaval that brought this about—the turmoil inevitably associated with the violent transition from a plurality of states to an empire. There was also a calculated element of cultural revolution. The Ch'in statesmen were guided by a single value: the maximization of the efficiency and power of the state. Whatever got in the way of this had to go. Hence their standardization of much that had previously been diverse in Chinese culture: weights, measures, law, script. But the Ch'in authorities also made it their business to stamp out any aspect of the inherited culture they judged hostile to the new-style state. One of their theoreticians put it this way: "In the state of an intelligent ruler, there is no literature of books and bamboo strips, but the law is the only doctrine; there are no sayings of the early kings, but the officials are the only models." A leading Ch'in statesman accordingly made the practical recommendation that "those who use the past to criticize the present should be put to death together with their relatives." In other words, the past in itself had neither authority nor value, and those who claimed otherwise were engaged in subversion. The state therefore set about burning books and, allegedly, burying scholars. The result of all this was that only a limited part of the heritage of pre-imperial China survived.

This ruthlessly negative attitude to the past did not last. Under the Han and later dynasties, the imperial state renounced its cultural radicalism, and what remained of antiquity was carefully collected and preserved. What was not given up was the project of ruling China through a unitary imperial order. As embodied in successive dynasties down the centuries, this imperial order was neither totalitarian nor even particularly intolerant. But its very existence tended to circumscribe the range of approved elite culture.

It is for these reasons that the Ch'in unification marks a key break in the cultural history of China. The centuries prior to the break were a period of remarkable cultural ferment. Two things helped to stimulate this. One was the existence of many small states rather than one big one—much as in the Gangetic plain in the formative period of Indian civilization. The other was that around the fifth century B.C. China entered the Iron Age. The effect of this was to undermine the aristocratic dominance that had marked Chinese society in the Bronze Age. Battles were no longer won by aristocrats with their chariots and expensive bronze armaments; when the Ch'in eventually prevailed, it was thanks to massed infantry increasingly equipped with iron weapons. In the interval the Chinese were unusually free to think.

This, then, was the period of the "hundred schools," when the followers of Confucius (d. 479 B.C.) were just one school among many. Culturally they were the conservatives, committed to preserving the elite political tradition of the late Bronze Age as best they could in a changed environment; but they were men of their own time in that they sought to maintain the relevance of their tradition by subjecting it to moralizing reinterpretation (we will get a taste of this when we come to the ancestor cult in the next section). Within the political elite the antithesis of Confucian thought was that of the Legalists, who were the inspiration of the Ch'in statesmen. Outside the political elite, with their roots among the craftsmen of the expanding cities, were the Mohists—antiaristocratic, moralistic, and puritanical (their principles included "economy in funerals" and "condemnation of music"). Their surviving writings show a concern with formal logic and rigorous scientific thought that was to have no place in imperial China. In antithesis to all of these were the Taoists (or Daoists), with their teasing, mystical, antirationalist philosophy of spontaneity and withdrawal from public life.

This cultural plurality of early China was never fully restored after the Ch'in unification. Instead, the question was now which school of thought would be established as the ideological partner of the imperial order. Despite the initial dominance of the Legalists under the Ch'in, it became clear under the succeeding Han dynasty, even as early as the second century B.C., that the answer was to be the Confucians. A Confucian education now became the key to entering the imperial bureaucracy, and by T'ang times this link was institutionalized through a formal examination system.

From time to time Confucian dominance might be threatened, as it was to an extent by the rise of Buddhism in the mid-first millennium A.D. as a religion combining services to rulers with mass appeal; but the Confucians turned the tables on the Buddhists, assisted by a dramatic onslaught on the monasteries and their wealth mounted by the emperor in 845. Several centuries later, under the rule of the Mongol Yüan dynasty, it looked for a while as if the Confucians might lose out altogether; but in the early fourteenth century the Mongol ruler restored the traditional examination system, so that officials were once again chosen for their proven mastery of the Confucian tradition. The major change that came out of these various commotions was the formation and increasing dominance of what is known as Neo-Confucianism, a reshaping of the old Confucian tradition to meet the spiritual and philosophical challenge of Buddhism.

The Northern Frontier and Late Imperial China

The passing role of the Mongols in this intellectual drama brings us to yet another theme that first becomes prominent in Chinese history with the Ch'in, though in this case the timing may be somewhat fortuitous. This is the menace of the barbarians on the northern frontier. The key fact about the territory to the north of China was its suitability for pastoral nomads. This meant that there was little scope for the Chinese to expand in this direction; to conquer a nomadic society is a thankless task for a settled state, and the lands that nomads inhabit are normally unattractive to peasants. It is only in recent centuries that Manchuria and Inner Mongolia have become a real part of China, and not just an imperial fringe. So there was no northern equivalent to the ancient expansion of China toward the south. Instead, China confronted repeated military threats in this northern zone from the late third century B.C. onward. In the period of disunity following the demise of the Han dynasty, dynasties of barbarian origin from the north were established within the frontiers of China. There was more of this in the tenth century, and again in the twelfth. In the thirteenth century came the Mongol invasion, when for the first but not last time northern nomads conquered the whole of China. The Ming dynasty represented a proud restoration of native Chinese rule, but in the seventeenth century the Ming succumbed to yet another pastoral people from the north, the Manchus; it was they who established China's last imperial dynasty, the Ch'ing.

This sounds like a rather disastrous record, and from an ethnic Chinese viewpoint it was so. But a crucial feature of such nomadic conquests of China, whether partial or complete, was that the new rulers came to terms with Chinese civilization, and in general did so sooner rather than later. The major exception was the Mongols, who alone had a geographical perspective that made them aware of alternatives. So the Mongols for a while curtailed the role of the native Chinese elite and patronized a motley array of Tibetan lamas, Muslim tax collectors, and other miscellaneous foreigners. The Chinese detested the Tibetan lamas for their arrogance and the Muslim tax collectors for their rapaciousness, and in due course were glad to see the last of the Mongols themselves; but in the meantime, as we have seen, the Mongols had come to terms with the Chinese elite. From a military point of view China's northern frontier was more dangerous than India's northwestern gateway; but culturally China never experienced anything like a Muslim invasion from that quarter.

Against this background let us glance at China under the Ch'ing, in the period between the Manchu conquest that established the dynasty in the mid-seventeenth century and the cataclysms that broke its power in the mid-nineteenth. The barbarian origin of the Manchu conquerors of China was always to some degree a problem, but particularly so at the beginning and the end of their rule: the dynasty was hated as alien by the Ming loyalists who resisted its advent, just as it was to be by the nationalists who finally overthrew it in the early twentieth century. In the face of such hostility, the Manchus took the remarkable step of obliging all adult male Chinese to manifest their allegiance to the new dynasty by adopting the distinctive Manchu hairstyle, the shaved pate and the pigtail or queue; as the Chinese put it, the choice was to "keep your hair and lose your head, or lose your hair and keep your head," and they acted accordingly. But seen from the other side of the street the Manchu problem looked very different: it was not that the Manchus were too foreign, rather that they soon became too Chinese. One Ch'ing emperor of the eighteenth century felt it necessary to make Manchus take examinations in the Manchu language, a clear sign of how far the rot had already spread—real barbarians do not take examinations, least of all to test their knowledge of their own language. Perhaps the only people who were thoroughly comfortable with the Manchu role in China were the Koreans; now that China was languishing under barbarian rule, they could flatter themselves that they had become the last repository of true Chinese culture. This did not mean that they were so imprudent as to refuse to recognize the overlordship of the Ch'ing; they duly sent their tribute missions to Peking, but might use the occasion to complain about derogatory references to Korea in Chinese history books.

Whatever the Koreans thought, the Chinese society over which the Ch'ing held sway was in many ways a thriving one. The Chinese population was growing substantially, and the economy grew with it, approaching an unprecedented size and complexity in all its major sectors—agricultural, industrial, and commercial. These trends were not new, but they reached their consummation under the Ch'ing. Thus by the end of the eighteenth century, Chinese society represented a larger accumulation of people and wealth than at any time in antiquity. This was true for the country as a whole and even more so for the region of the Yangtze delta; in economic and social terms this part of the country had become its unchallenged center of gravity, despite the fact that the political capital was located far to the north in Peking. Yet, despite this growth, Chinese society had not changed in its fundamental character. It continued to be dominated by a landed elite, and the crucial interfaces were between the gentry and the peasantry, on the one hand, and between the gentry and the state, on the other. Though the economic expansion meant the existence of a substantial commercial bourgeoisie, its power was not commensurate with its wealth.

The picture was similar with regard to culture: there was more of it, and it was more sophisticated than ever, above all in the Yangtze delta region. For example, more studies of local history were written and printed than ever before; there were more novels to read in colloquial Chinese, more theaters played to the public, and so on. The prevalent intellectual trends remained broadly Neo-Confucian, in the tradition of the philosopher Chu Hsi (or Zhu Xi, d. 1200), but this did not mean an absence of significant innovations and commotions. The sixteenth century saw the birth of a new classicism that denounced the prose style of the Sung period and sought to return to ancient models; in later centuries this movement was rather forgotten in China, but it was to have a considerable impact in eighteenth-century Japan. A more lasting development was the emergence of a school of exact textual scholarship of a kind that had not previously existed in China. It was a seventeenth-century philologist who established that almost half of a revered Chinese classic, the *Book of Documents*, was a later forgery. One eighteenth-century scholar in this tradition strayed into philosophy, where he rejected the Neo-Confucian synthesis altogether; but he was an isolated figure. Moreover, the relaxed, if superficial, openness to European learning that had developed with the arrival of the Jesuits in the late Ming period was now less in evidence. It did, however, have a curious repercussion among the Chinese Muslims: they learned from the Jesuits how to defend a monotheist religion against the charge of being incompatible with Confucianism. A seventeenth-century Muslim scholar in Yunnan in the far southwest wrote a book in this vein; he even made a journey to Peking in an effort to obtain for the descendants of the Muslim prophet the same prestigious status that had traditionally been accorded to those of Confucius. He failed: imperial China tolerated diversity, but had better things to do than to celebrate it.

All in all, China under the Ch'ing dynasty is an instructive phenomenon in the history of Eurasia. It shows how far a traditional agrarian civilization could go without becoming something else.

Chinese Culture beyond the Imperial Frontiers

At several points in our survey of late imperial China, we have come up against the adoption of Chinese culture among non-Chinese peoples beyond the frontiers. Chinese influence, of course, was widespread in East Asia, but instances of the lasting reception of an integral Chinese culture by the non-Chinese elites of independent states were relatively rare. The north, despite some interesting developments, was not congenial territory for such a process, and in the west and south, as we saw in the preceding chapter, those who had the alternative of choosing Indian culture usually took it—perhaps because the shape of Chinese culture required the entire elite to become learned, whereas the shape of Indian culture meant that it was enough for them to import and patronize the learning of monks or Brahmins. So it is not surprising that two of the three cases that concern us were located in the settled, agricultural lands of the far northeast, one being the Korean Peninsula, and the other its island neighbor Japan. Here geography ensured that neither country was exposed to Indian culture except through China, and it was by this route that they adopted Buddhism. In both these cases the key period for the large-scale adoption of Chinese culture was around the middle of the first millennium A.D.; in the Japanese case Koreans played a major role as intermediaries (though this was sometimes conveniently overlooked, as was already pointed out by the ninth-century Shinto priest Imbe no Hironari). The third case was Vietnam, the corner of Southeast Asia farthest from India and closest to China. (The southward expansion of Vietnam took place only in recent centuries, and in the course of it the Vietnamese overran regions that had previously adopted Indian culture.) In Vietnam the key period was perhaps half a millennium later than in the northeast. Of the three countries it was Korea that took the adoption of Chinese culture furthest. Here the fifteenth century saw a radical attempt by the state to Confucianize Korean kinship structures, and by the early seventeenth century a Manchu ruler could comment that the Chinese and Koreans differed only in language.

Yet it was in Japan that the interaction of Chinese culture with the native society had the most idiosyncratic and historically significant results. There is no denying that Chinese influence went very deep—so deep that the basic Japanese number system today is made up of Chinese loanwords. In the seventh to the ninth century the Japanese imperial institution was in many ways a provincial replica of that of the T'ang dynasty; Japanese legal texts of this period have been used by historians to reconstruct T'ang legislation that has gone lost in China. But two things made the Japanese evolution very different from that of Korea in relation to China.

One was political. Instead of a succession of Chinese-style dynasties, the Japanese developed a curious dualism: while the original imperial dynasty went on forever, it lost all but the trappings of power, and the reality of power passed into the hands of military rulers who came to be called shoguns. The insularity of Japan does something to explain this: being somewhat sheltered from invasion, islands are less subject to the harsh continental disciplines that drive the formation of unitary states and sweep away obsolete institutions. Before the nineteenth century there was only one, short period when an emperor sought to recover his ancient powers, the abortive Kemmu Restoration of 1333–36. Its most emotive figure, at least in Japanese historical retrospect, was the ill-fated restorationist leader Kusunoki Masashige. In fact, the retrospect was to prove more important than the events themselves.

The other divergence between the paths of Japan and Korea was cultural. As the language of literature, Chinese retained an undisputed hegemony in Korea until modern times. It was likewise a prestigious literary language in Japan; but to a much greater extent than the Koreans, the Japanese took to "reading" it through a process of mechanical translation into their native language. Doubtless following Korean precedent, the Japanese adapted the Chinese script to write Japanese at an early date; but the corpus of literature they composed in their own language was far larger than the analogous literature of Korea. Its most famous product was the *Tale of Genji*, a romance written in the early eleventh century by one of several female authors of the period. In sum, despite massive Chinese influence, Japan was significantly more Japanese than Korea was Korean. It was also an intellectually more plural society: it was in Korea, not Japan, that a leading seventeenth-century scholar held that "a man who does not believe in Chu Hsi is a barbarian."

II. Keeping in Touch with the Ancestors

In the second half of the ninth century A.D., the son of a saintly Muslim accepted from the caliph an appointment as a judge in the Iranian city of Isfahan. Holding this office meant depending on the caliph for a living. It also involved dressing in black, the color associated with the reigning dynasty. For someone from a pious background, it did not feel good to be on the dynasty's payroll and wear its color; debt and too large a family had forced him to take the job. In fact, he felt so bad about it that when he arrived in Isfahan, he wept in public at the thought of his father seeing him in such circumstances. Now, in real life there could have been no question of his father seeing him, since he was already dead; it was the thought that counted. There is nothing specifically Muslim about such a thought, which might occur to any of us, for better or worse: "What would my father (or grandfather, or great-grandfather) say if he could see me now?" But it would not occur to most of us, Muslims and others, to elaborate this simple thought into a formal ritual. Yet in China this takes us to the heart of a practice as characteristic of the culture as caste is of India—the ancestor cult.

The Neo-Confucian philosopher Chu Hsi wrote a short guide to the performance of the standard Chinese family rituals, partly in order to discourage recourse to Buddhist ceremonies. His book became very influential in China (as also in Korea—though not, significantly, in Japan). One of the subjects with which he dealt was reporting events to the ancestors. Suppose, for example, that like our unhappy Muslim you have been appointed to office. You respectfully visit the offering hall where the tablets of your ancestors are arranged, and follow a standard ritual. After tea and wine have been offered to the spirits of the ancestors, you have the report read out from a prayer board. It should announce to the ancestors that on such-and-such a date "your filial son" was appointed to such-and-such a position. "Due to the teachings of his ancestors, he now enjoys rank and salary. For the benefits he has received, he is overcome by gratitude and admiration. Earnestly, with wine and fruit, he extends this devout report." Of course, the news is not always so good, and instead you may find yourself having to reveal that you have been demoted or dismissed. Here a somewhat different tone is in order: "having discarded the ancestral teachings, one is in trepidation and uneasy. Earnestly, etc."

One thing that emerges here is a marked divergence of attitude toward holding a position in government. It was receiving such a position, not losing it, that filled our Muslim judge with trepidation and unease. Traditional Chinese attitudes were very different, and this also comes out in other aspects of the rituals of the ancestor cult. Naturally you would not visit the offering hall to make a report unless suitably dressed; if you are in office, this means that you wear your official robes and hold your official plaque. (Our Muslim judge, by contrast, made a habit of taking off his official dress after leaving his court.) Even your ancestors may have a continuing stake in official life. You could be fortunate enough to have the honor of reporting titles and offices conferred on them after their deaths, in which case their tablets should be reinscribed accordingly. Confucius, for example, was honored with substantial posthumous promotions under the T'ang dynasty; this doubtless pleased his descendants, some of whom were quite prominent at the time. The state, in short, is taken very seriously as a moral agent. Indeed, Chu Hsi modestly informs us in his short preface that one of his purposes in writing his guide was to "make a small contribution to the state's effort to transform and lead the people."

There is another point of interest in this preface, namely the attitude it reveals toward the Confucian classics that had been handed down from the Chou period. In those ancient times, Chu Hsi tells us, the classical texts on ritual were entirely adequate. Yet the regulations and instructions in the surviving texts are "no longer suited to our age." In adapting what they say, he emphasizes, one must achieve a "proper balance"; otherwise one may end up concentrating on secondary elements and forgetting those that really matter. He tells us that he therefore went about things by first identifying the major structures that cannot be changed, and then making "minor emendations." In other words, his attitude to the details of the classical texts is flexible, in recognition of the fact that times have changed. As a fully committed Confucian, he follows Confucius' idea of "carrying on what came from our predecessors"; but he is far from being a fundamentalist. In later centuries he was criticized with regard to the specific ways in which he had departed from the classics, but not on the grounds that one should not depart from them at all; in this respect his overall position is fully representative of the Confucian mainstream. Tu Yu (or Du Yu, d. 812), a Confucian of the T'ang period, had expressed a similar view: "Whenever one consults the books of the ancients, it is because one wishes to reveal new meanings and form institutions in accordance with present circumstances. Their way is inexhaustible. How much more are plans for contrivances and expedients subject to a thousand changes and ten thousand alterations. If one imitates in detail, it is like notching a boat to mark a spot."

A final point about the preface is a certain metaphysical silence. Christians split hairs trying to determine the precise sense in which Jesus was or was not God (is it admissible to say that God had a mother and died on the cross?). Muslims and Buddhists have comparable doctrinal concerns. Yet Chu Hsi, philosopher though he is, raises no questions about the ancestor cult in this vein. He says that he has given "a high place to love and respect," which he considers to be among "the fundamental elements." But he shows no interest in posing the question of the exact sense in which the spirits of the ancestors can be said to be present in the offering hall and to understand what is reported to them; we are left to suspect that the answer may not really matter. It is not that Chu Hsi was a skeptic: from remarks he makes elsewhere, we know that he believed that ancestral spirits had some kind of existence and that there could be real contact with them in the course of the rituals. But it is not a question he is interested in exploring: "This," he tells his students to shut them up, "is a matter difficult to talk about, so I simply ask that you think about it for yourselves." In any case, the straightforward bargain underlying the early Chinese ancestor cult—food for blessings—is scarcely to be detected in his rituals. We find nothing there to compare with the words of an ancient poem: "The spirits enjoyed their drink and food, and will grant our lord a long life." The rites Chu Hsi sets out are in this respect significantly different from Vedic rituals. The latter are designed to get results by influencing the behavior of the gods; by contrast, how to prod the ancestors into delivering goods is not a central concern of Chu Hsi's. This was appropriate for a Confucian. Confucius had advised that one should be reverent to ghosts and gods, but keep them at a distance. Such ambivalence irritated the Mohists, who liked things cut-and-dried: it was self-contradictory, their founder complained, for a Confucian to affirm that "gods and ghosts do not exist" while maintaining that "the gentleman must learn the sacrificial ceremonies"—like making a fishnet though there are no fish. But if there was a contradiction here, the Confucians seem to have lived with it quite comfortably.

What is at stake in the ancestor cult is not the minutiae of ancient rituals, or the metaphysics of ghostliness, but family values (and more broadly, human relations). Confucius says, "When one's parents are alive, one serves them in accordance with the rites; when they are dead, one buries them in accordance with the rites and sacrifices to them in accordance with the rites." The great moral touchstone is filial piety: "When a person's father is alive, observe his intentions. After his father is no more, observe his actions. If for three years he does not change his father's ways, he is worthy to be called filial." Confucianism, it might be said, is the true ethic of family values. Jesus, by contrast, told the crowds traveling with him: "Whoever comes to me and does not hate father and mother, wife and children, brothers and sisters, yes, and even life itself, cannot be my disciple" (Luke 14:25–26). A man he instructed to follow him replied, "Lord, first let me go and bury my father." Yet Jesus told him, "Let the dead bury their own dead; but as for you, go and proclaim the kingdom of God" (Luke 9:59–60). The kingdom of God is the key to this exhortation to filial impiety: Jesus had something to say that in his view was far more important than family values. The same was true of the Buddha; small wonder that a standard Chinese criticism of Buddhism was that it incited men to abandon their parents by becoming monks. Both Jesus and the Buddha proclaimed a message that sought to jerk people loose from the social nexus into which they had been born. Confucianism, by contrast, is about keeping faith with that nexus. And in any case Confucius brought nothing so overstated as a message. He once said, "I should like to do without speech." One of his disciples objected, "If you do not speak, what message will your disciples have from you?" To this Confucius replied, "Does Heaven speak? The four seasons proceed by it, the hundred things are generated by it. Does Heaven speak?"

There was, of course, much more to the Chinese ancestor cult than can be gleaned from the sayings of Confucius and the prescriptions of Chu Hsi. It was already present in an elaborate form in the ritual life of the late Shang kings as documented in the oracle bones, and such dynastic forms of the cult remained prominent in imperial China. Statistics of the first century B.C. reveal the care of the imperial ancestors as a veritable pork barrel: it involved 343 shrines, with 45,129 guardians and 12,147 priests, cooks, and musicians; the dynastic ancestors were being served a total of 24,455 meals a year. Understandably it took savage cuts in ancestral welfare to balance the budget. But the cult also permeated Chinese society, in forms that might approximate or diverge from those enjoined by Chu Hsi in all sorts of ways. This, after all, was a society that greatly respected elders; ancestors were even older. When the Jesuits applied themselves to converting the Chinese to Christianity in the sixteenth and seventeenth centuries, a major issue they confronted was what to make of the ancestor cult.

By now we are used to such regional cultural idiosyncrasies as the Chinese ancestor cult. We regularly ask why they are there, and regularly remain in the dark. As with caste in India, we can look to geography for a minor part of the explanation. Given that China was cut off from dense linkages with other societies on all sides but the north, and that its direct northern contacts were with pastoral nomads rather than with another civilization, it is not a puzzle that Chinese society should have developed along rather distinctive lines. But this does nothing to explain why it was the ancestor cult that became such a prominent and pervasive feature of the society.

III. Shang Bronzes: What's in a Name?

Figure 16 shows a bronze vessel—specifically a *chia* (or *jia*)—dating from the fifteenth or fourteenth century B.C. As the sketches accompanying it suggest, this particular type is one of a family; in fact, many more varieties of bronze vessel were manufactured than are represented here. Large museums are full of them. The *chia*, however, is relatively easy to distinguish, thanks to the pair of mushroom-like protuberances from the rim. What, then, was its significance?

At one level the answer is easy. The *chia* and its relatives were adjuncts of an aristocratic way of life to which enormous resources were devoted in China for something like a thousand years. As was mentioned above, bronze artifacts had appeared in the far northwest of China around 2000 B.C., though there is no indication that their role at that time and place was tied to the existence of an aristocracy. But within a few centuries bronze was being made farther to the east, in the part of the Yellow River valley associated with the Shang. Here the whole character of the enterprise was different. Taking advantage of China's unusually rich mineral resources, this society was producing bronze artifacts on a scale unmatched anywhere in the world of the day; many of these were vessels like those shown in figure 16 (see pages 180–81). At the same time they were employing an unusually extravagant technology: instead of working their bronze by hammering it in thin sheets, as was the normal practice in cost-conscious Bronze Age societies, they were casting their vessels in clay molds. Their generosity was also evident in the sheer mass of metal they were prepared to bury with their dead. A Shang queen was interred with 195 bronze vessels, and the total number recovered from tombs of the Shang period must run into thousands. This scale of activity presupposes a strongly stratified society, one in which the labor of the common people was being mobilized relentlessly to sustain the aristocratic lifestyle of the elite.

But what was the exact role of the *chia* in this way of life? The contents of the tombs indicate that bronze vessels came in sets, suggesting that each had its particular purpose. It is also known from inscriptions on some vessels that they played a part in the ancestor cult. It seems that one class of vessels, including the *chia*, was for offerings of wine, while another class, including the *ting* (or *ding*), was for food. A *ting* containing animal bones has indeed been found in a rather late burial, but we have no such direct confirmation for the *chia*. If it actually was used for wine—to heat it or pour it—how, then, did its function differ from that of other vessels in the same class? We have no contemporary accounts to help us here; and we cannot look to later tradition for reliable information, since the *chia* (unlike the *ting*) went out of use early in the Chou period.

If we know so little about the precise ritual function of this vessel, how do we even know that it was called a *chia*? The answer, unsurprisingly, is that we don't. The word *chia* is authentic enough (though we can have little idea how it was pronounced in Shang times). But there is no good reason to think that the word was applied by contemporaries to the type of vessel with which we now associate it. We might therefore suspect a modern misnomer, a sin of which archaeologists are quite often guilty. Faced with an object for which they lack a technical term, they have a tendency to help themselves to a word from the relevant culture and misapply it. Thus modern students of the pottery of ancient Greece use an array of indigenous terms in ways that a native speaker of ancient Greek might have found bizarre. The *chia* constitutes an analogous case. What makes it interesting is that the culprit was not a modern archaeologist: it was a Chinese antiquarian of the eleventh century A.D. who gave *chia* its current meaning.

Long before Sung times the Chinese had begun to show a lively interest in antiquities. Bronze vessels, especially *ting*s, played a prominent part in this. As early as 113 B.C. the discovery of an ancient and precious *ting* caused a sensation at the Han court. In later centuries rulers and others set about collecting on a large scale, much as they had done in the last decades of ancient Mesopotamian kingship. But this interest was not in general a scholarly one; thus the commentary provided by the court officials in 113 B.C. did not go beyond mythological claptrap. In effect these objects were valued more as talismans than as antiques. We see the other side of the coin in an event of A.D. 591, when some ancient vessels acquired by the new Sui dynasty from a defeated enemy were destroyed; the reason given for this drastic step was that the vessels were found to possess malign magical properties. Such attitudes were remarkably persistent. The author of a rather serious handbook for collectors published in 1388 remarks that ancient bronze vessels can often ward off evil spirits and should accordingly be kept in the home.

But by Sung times the leading scholars were already setting a rather different tone. They still used a rhetoric in which the objects of the past provided models for the present, but what seems to have excited them most was the philological challenge of the ancient inscriptions sometimes found on the vessels. Here, as one scholar pointed out, were contemporary sources that could be used to correct errors in the history books. One antiquarian who searched for ancient bronze vessels had pictures of them engraved, and distributed rubbings of the inscriptions to his fellow scholars (there was a brisk market for rubbings at the time—merchants would buy them in the north and sell them in the south). Another scholar published a ten-volume catalog of bronzes in 1092, drawing on no fewer than sixty-five collections. Figure 17 shows his entry on a *ting* found at the site of the late Shang capital, taken from the 1229 printing of his work—the earliest printing of which copies survive today (in China printing goes back at least

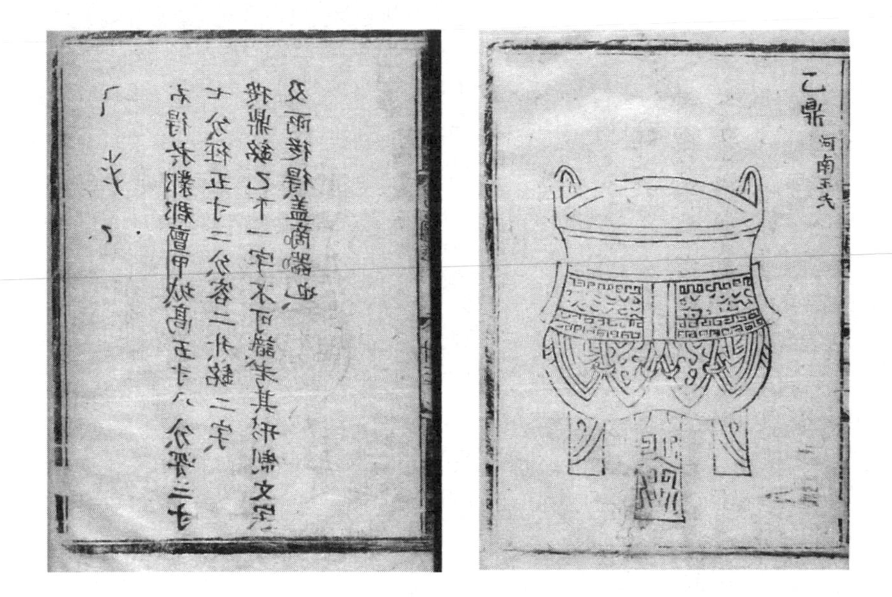

Fig. 17: An entry on a *ting* from a Sung catalog of bronzes.

to the seventh century). Chu Hsi himself wrote a short essay called "Our family collection of ancient inscriptions."

After Sung times there was a lack of new scholarship for several centuries, though the old works were often reprinted and collecting seems to have continued unabated. But under the Ch'ing dynasty the serious study of ancient bronzes revived, and a catalog published by imperial command in 1755 ran to forty volumes. By this time, however, the interest of the Chinese elite in antiquities had engendered a massive revenge effect: many of the fifteen hundred bronze vessels included in the new catalog were forgeries of the Ming period. It is in fact rather surprising that the authors of the catalog had not done a better job of weeding out the spurious vessels, for the author of the handbook of 1388 was already paying attention to the problem. He describes the methods that were being used to make new bronze look old, and remarks that vessels showing traces of the use of the chisel are fakes. Be this as it may, China was one of the few parts of the world where the market for antiques was already brisk enough to elicit large-scale forgery in premodern times. As early as the fifth century A.D. we have an account of the methods that were in use in China to fake old specimens of calligraphy. This was not unique: five hundred years earlier, fake artworks bearing spurious signatures of classical Greek masters sold for high prices in ancient Rome. But nothing like that has been detected in ancient Mesopotamia.

CHAPTER 10

THE ANCIENT MEDITERRANEAN WORLD

I. Frogs around a Pool

As we saw in chapter 4, there was once a mid-world ocean separating the two supercontinents of Laurasia and Gondwana. Along much of the southern fringe of Eurasia this ocean has disappeared: the various fragments of Gondwana have either crashed into Eurasia and joined it or drifted away. But there is one major exception to this pattern, namely the region bounded by the western third of Eurasia to the north and by Africa to the south. Here a segment of the old mid-world ocean has been preserved (though it has dried up in the geological past when cut off at the Straits of Gibraltar). The eastern and western regions of the Old World are thus in sharp contrast: whereas in the east the central geographical feature is the massive Tibetan plateau, in the west it is the Mediterranean Sea.

This account of the Mediterranean as a survival from an earlier epoch is, of course, a simplification. Eurasia and Africa have not simply stood still since the opening of the mid-world ocean; as little as ten million years ago the Mediterranean looked distinctly different from the way it does today. What we now confront is the outcome of a complex, not to say chaotic, geological evolution. In some places continental crust has collapsed; in others fragments of it have been scattered through the sea as islands or peninsulas. Collisional effects have led to much mountain building in geologically recent times, and high Holocene sea levels have made further changes to the map. In addition to all this, none of these processes has operated in a geographically symmetrical way.

The simplest way to sum up the results is to compare the physical geography of the northern and southern shores of the Mediterranean (we can leave aside the Syrian coast, which closes off the sea on the east). On the north are four major peninsulas: Spain (alias the Iberian Peninsula), Italy, Greece, and Anatolia. Italy and Greece are entirely contained within the Mediterranean, whereas Spain and Anatolia enjoy peninsular status by virtue of other adjoining bodies of water: in the case of Spain the Atlantic Ocean, and in the case of Anatolia the Black Sea (though for some time before the early Holocene it seems to have been no more than a freshwater lake). All four peninsulas are mountainous, but Spain and Anatolia are distinguished from the purely Mediterranean ones by their possession of large interior plateaus. The southern shore, by contrast, lacks comparable peninsulas, and while it is mountainous in the west, most of it is rather flat.

We have already encountered the Mediterranean climate in the context of the Near East. Thanks to the east–west orientation of the Mediterranean, it is climatically rather homogeneous; the overall pattern is hot, dry summers and mild, wet—though not very wet—winters. But the winter rainfall is unevenly distributed (see map 8). As the location of the mountains would lead us to expect, the north is more fortunate than the south, and the southwest is better served than the southeast. Large rivers can, of course, bring water to arid areas; but the only major instance of this in the Mediterranean world is the Nile.

What did all this mean for the potential of the Mediterranean world as an environment for humans? It clearly made it very different from the regions to the east, and that in two main ways. On land there was nothing outside Egypt to compare with the great river valleys of the Near East, India, and China; Mediterranean agriculture was accordingly practiced on small, scattered plains, or failing that in the hills, and moving around the region could be slow and painful. But at sea it was a different story. The presence of humans on a few of the islands in Mesolithic times shows that even before the onset of the Neolithic some kind of seafaring had developed. Once seaworthy ships came into use, the existence of the Mediterranean made possible a precociously interconnected world (like "frogs around a pool," as Plato put it); and the shared environmental conditions of this world meant that what worked well in one part of it was also likely to work well in another. To the east, by contrast, it was much harder to circumvent the Tibetan plateau or cope with the vastness of the Indian Ocean, and geographical conditions along the way were far more varied.

Despite the relative homogeneity of the Mediterrancan world, it was significant for its history that the north tended to be more favored than the south. Egypt apart, the distribution of rainfall meant that the agricultural resources of the northern shore were in general better, or less bad, than those of the southern shore. The north was also better endowed from a maritime point of view: its coastline was more indented and thus more friendly to early shipping, and its rainfall provided it with more timber to build its ships. On this latter score Egypt shared fully in the disadvantage of the south.

Mediterranean Prehistory

Although we did not look at the Mediterranean as a whole in earlier chapters, we did encounter large parts of it in connection with Africa and the Near East. The main focus of this chapter will

Irish

ATLANTIC
OCEAN

0 200 miles
0 200 kilometers

Galicians
Basques
PYRENEES
G A U L
A L P S
Venice
S P A I N
Salamanca
PORTUGAL
ARAGON
Toledo
IBERIAN
PENINSULA
ANDALUSIA
Etruscans
Rome
I T A L Y
Carthage
Sicily
Berbers
N O R T H A F R I C A
Berbers

MAP 8: THE MEDITERRANEAN WORLD
Modern boundaries
RAINFALL (cm/yr)
More than 100
50–100
30–49
20–29
Less than 20
Goths
Black Sea
MACEDONIA
THRACE
GREECE
Aegean Sea
ATTICA
Athens
Mycenaeans
Sparta
Crete
Minoans
Mediterranean Sea
Hittites
LYDIA
ANATOLIA
(TURKEY)
CARIA
Cyprus
SYRIA
NEAR EAST
PHOENICIA
Tyre
Israelites
Jerusalem
Cyrenaica
Berbers
EGYPT
Nile River

accordingly be on the European sector of the Mediterranean coastlands. But what happens in this region has to be seen in the wider context. It is, for example, a key fact about the prehistory of the Mediterranean world that it includes within itself the western margin of the Near East, not to mention Egypt.

The role of the Near East is already evident in the Upper Palaeolithic. Archaeology suggests that modern humans appeared in Europe about forty thousand years ago and that their arrival was followed by the extinction of the indigenous Neanderthal population within the next ten thousand years. The genetic evidence confirms this picture and points to a parallel occupation of northern Africa by a related population in roughly the same period. It further indicates that both these modern human populations originated in the Near East—which is very much what one would expect in the case of Europe, though not perhaps in that of northern Africa.

It is likewise clear that farming spread to Europe from the Near East, just as it seems to have done in northern Africa. Several of the domesticated species involved could only have been of Near Eastern origin. Moreover, the relative dates at which they appear point to an expansion from the Near East; thus we already find farming in Greece in the eighth millennium B.C., whereas it does not appear in Spain until the sixth. But by 5000 B.C. farming was firmly established over most of the Mediterranean world. What has recently become controversial is the process by which the spread took place: Was farming brought to Europe by a population of Near Eastern origin, or did it expand across Europe through adoption by indigenous hunter-gatherers? For a generation it was widely accepted that migration was the dominant process, but currently the genetic evidence indicates that the population of Europe today is mainly of Palaeolithic origin. A broadly similar pattern of diffusion is likely to hold for the spread of metalworking. The northern shore of the Mediterranean had entered the Bronze Age by the end of the fourth millennium in Greece and by the end of the third in Spain; likewise Greece entered the Iron Age in the eleventh century and Italy around the eighth.

The earliest appearance of civilization in Mediterranean Europe again points eastward. About 2000 B.C. Minoan civilization emerged on the island of Crete; writing appears around the eighteenth century. A century or so after that the closely related Mycenaean culture appeared on the Greek mainland, but as yet no comparable culture is in evidence farther west. This tradition of civilization was not a direct local adoption of any one Near Eastern culture. For example, both the Minoans and the Mycenaeans wrote on clay tablets, but the scripts they used, which were obviously related, have no known Near Eastern model. Yet it can hardly be an accident that these civilizations made their appearance in the part of Mediterranean Europe closest to Egypt and the Near East, and in fact the chronology of Minoan civilization turns on Egyptian and Near Eastern artifacts found in association with its pottery.

This first European civilization is known mainly from the ruins of palaces ruling what must have been rather small states, and from the limited corpus of administrative records found in them. In the Mycenaean case these records have been deciphered sufficiently to establish one key point: that they are written in an early form of Greek, the language that has been spoken in the region ever since. Greek is an Indo-European language, and is likely to have been brought to Greece at some point in the Bronze Age—though just when is a matter of guesswork. Since the language of the earlier Minoan tablets seems not to be Greek, the presumption is that Mycenaean culture represents a Greek adoption of that of the non-Greek Minoans. Toward the end of the second millennium B.C. this whole tradition of civilization collapsed; and when civilization reappeared in Mediterranean Europe in the first millennium, it represented a fresh start. Only on Cyprus did a script of Mycenaean origin survive, perhaps brought there by refugees, and this was to prove a dead end. There is an obvious parallel here to the demise of India's first civilization. But at least two things were different: there was linguistic continuity across the divide, and a memory of the Mycenaean world survived, however hazily, to find its place in the Homeric epics.

Early Mediterranean History

The new civilization that appeared in the first millennium B.C. goes back in at least one key respect to the Phoenicians, who were perhaps the first people to establish a maritime presence that extended over the entire length of the Mediterranean. We encountered their activities in chapter 6 in connection with their cultural impact on northwestern Africa. In their homeland on the central Syrian coast they lived in a plurality of small states, and for the most part the establishment of modest coastal colonies in other parts of the Mediterranean was the limit of their territorial ambitions. They were not empire builders, nor were they hungry for arable land. Their business consisted in trade with the indigenous peoples.

One such people was the Greeks, and it was from the Phoenicians that in—or by—the eighth century B.C. the Greeks derived the alphabetic script that they use to this day. As we noted in chapter 3, the Greeks transmitted the alphabet to peoples farther west, to regions in which any form of writing had previously been unknown. Thus in the seventh century a form of the Greek alphabet was being used by the Etruscans in central Italy. They in turn transmitted it to some minor Italian peoples who by the sixth century included the Romans. A little later a script of unclear, perhaps mixed origin came into use among the indigenous population of Spain; and as we saw in chapter 6, the native population of North Africa eventually adopted the Phoenician script. By now writing was established all around the Mediterranean, though outside the east only the Greeks and the Romans produced literary heritages that survived the vicissitudes of history.

One thing this epidemic of writing reveals, mostly for the first time, is the ethnic makeup of the Mediterranean world. As an environment in which to develop ethnic homogeneity, a peninsula is almost as good as an island; but by and large this potential had yet to be realized in the first millennium B.C. In the east Anatolia was inhabited by numerous distinct peoples speaking different languages, and the Hittite records show that this had also been true in the second millennium. To the west the much narrower Italian peninsula was almost as varied. In the far west of the Mediterranean the indigenous population of Spain was less heterogeneous, but there was still a plurality of peoples and languages (one of them doubtless ancestral to Basque). In each of these peninsulas at least one of the languages was Indo-European and at least one was not. A situation of this kind may also have characterized Greece in the second millennium: Greek was Indo-European, whereas Minoan clearly was not; and on the mainland the Greeks later remembered having shared their country with a people they called the Pelasgians. But in the first millennium non-Greek peoples and languages were no more than a residual presence, and Greek itself was a single language divided only into dialects. The Greeks thus possessed an unusual ethnic and linguistic homogeneity, and they were conscious of it. They called themselves Hellenes, saw themselves as a people sharing the same ancestry, language, and customs, and even possessed a few pan-Hellenic institutions like the Delphic Oracle and the Olympic Games.

Mediterranean Politics

Another feature of the Mediterranean scene that written sources illuminate is the political organization of its peoples. Here we find a striking gradient as we move from east to west. The Hittite state in central Anatolia was a sizable one by the standards of the second millennium, and a serious contender in the political and military affairs of the Near East; in the next millennium the Lydian state came to rule most of western Anatolia, until in the middle of the sixth century it was conquered by the Persians and became part of an even larger state. But nowhere west of Anatolia was anything to be found on this scale. The Mycenaeans had their states, if we can judge by their palaces and the later epic tradition, but they were small; and when Greece comes back into focus a few centuries later, the scale of political organization is, if anything, smaller. The same lack of unity marks Italy for most of the first millennium. Meanwhile in Spain, as in northern Africa west of Egypt, we know of nothing that deserves the name of a state. In comparative terms, all this is rather breathtaking. On the basis of our survey of the Old World civilizations so far, we would hazard the generalization that civilized people tend to produce a reasonable number of states of a certain size: Egypt, Magadha, Ch'u, and the like. So in this respect at least, the central and western Mediterranean was a region of unusual backwardness. Its extreme political fragmentation clearly had much to do with its refractory terrain and the limitations of its agricultural resources, though as we shall soon see these did not preclude the formation of larger states in later times.

Just as significant as the degree of fragmentation is the nature of the fragments. In the eastern Mediterranean we regularly have to do with powerful kingdoms, as in the cases of Lydia and Egypt; the Mycenaean states were perhaps similar structures on a smaller scale. At the other end of the Mediterranean, in Spain as in most of northern Africa, we dimly perceive tribes and chiefdoms. But the single most characteristic feature of the political organization of the Mediterranean in this period is the salience of the city-state.

City-states were not unknown in the ancient Near East; indeed, they were present in Mesopotamia at the dawn of history. But more directly relevant to our concerns is the emergence of the Phoenician city-states in the second half of the second millennium. As we saw in chapter 6, in at least one instance the Phoenicians replicated the city-state in the western Mediterranean, namely at Carthage. But the institution was more extensively adopted, perhaps under Phoenician influence, by the native peoples of Greece and Italy. The Greeks in turn spread it by establishing numerous colonies overseas, on the coasts of Anatolia,

Cyrenaica, southern Italy, southern France, and even northeastern Spain. The Mediterranean world thus came to be dotted with independent city-states. These were relatively easy to establish and maintain in regions where the hinterland lacked effective political organization, as was long the case in the western Mediterranean. In the eastern Mediterranean, by contrast, the political environment was less favorable. Along the Egyptian coast city-states simply failed to appear, and the Egyptian ruler funneled Greek trade through a single emporium. In other regions they lost their independence with the rise of effective states in the hinterland. Such was the fate of the Greek settlements of western Anatolia with the rise of Lydia, and its subsequent conquest by the Persians. Meanwhile, the cities of Phoenicia itself had to come to terms with a whole series of overlords.

City-states were often monarchies, or at least started that way. The Phoenician cities, for example, had their kings, as did the Etruscan cities, though these monarchs might not be very imposing; a Roman assassin failed to kill an Etruscan king because he could not tell the difference between the king and his secretary. But in general, monarchic city-states were not simply petty kingdoms. In the Phoenician cities, despite the real power wielded by the kings, we also hear of councils of elders, in one case with a membership of a hundred; thus there were formal institutions that gave some standing in the political process to the leading citizens, if not to the citizens at large. One city, Tyre, was for a while a republic ruled by a pair of judges. Unfortunately we do not know enough about Phoenician politics to say much more than this. But institutions of the same kind turn up in city-states elsewhere in the Mediterranean, and in some cases we have considerable knowledge of the rules under which they functioned. The information is naturally richest for the city-states at the center of the only two literary heritages that survive from the ancient Mediterranean world outside the Near East, that is to say for Athens and Rome. But we also have quite detailed accounts of the constitutions of such cities as Sparta and Carthage. As we will see in the next section, it was in the context of the constitutional arrangements of city-states that democracy first made its appearance.

An obvious weakness of city-states, and one we have already noted, was that they could not easily stand up to the power of large territorial states. A single city-state was usually too small to resist effectively, and rivalries between cities tended to prevent or disrupt alliances. Since the Mediterranean had considerably more potential to support large states than had yet been realized, this meant that the world of city-states was in the long run unsustainable. Either large states would move in and take over, or one city-state would expand and subject its peers.

What happened in Greece was on balance an illustration of the first process. In the early fifth century B.C. the country was invaded by the Persian Empire, but the Greek resistance proved successful, thanks to an alliance led by the two most powerful city-states, Athens and Sparta. The Athenian role in this alliance in turn laid the foundations for an Athenian maritime hegemony in Greece; but this was broken by a war between Athens and Sparta in the last decades of the fifth century. The Greek city-states had thus survived both an external and an internal threat to their traditional political way of life. But the fourth century saw the rise of a new external threat in the form of the Macedonian kingdom to the north, and by 338 B.C. a Macedonian hegemony had been established over Greece that sharply curtailed the political activity of the city-states. The end came with the incorporation of Greece into the expanding empire of the Romans in the second century B.C. It was thus the formation of large states on the edges of the Greek world that spelled doom for the city-states of Greece. The puzzle is perhaps that the process should have taken so long to complete.

In Italy, by contrast, it was the rise of one city-state that ended the independence of the others. Rome began its life on the edges of the Etruscan world, ruled by a dynasty of Etruscan kings. At a date traditionally given as 510 B.C., the Romans expelled the royal family and instituted a republic. In the centuries that followed they displayed an unusual combination of militaristic aggressiveness and willingness to incorporate defeated enemies into the enterprise of further expansion. They established their hegemony over Italy and went on to defeat Carthage, their chief rival in the western Mediterranean, thereby bringing the whole of this region under their direct or indirect rule. They then extended their sway to the eastern Mediterranean, completing the entire process in the first century B.C. A world of independent city-states had become a pan-Mediterranean empire.

The contrast in scale between these two forms of political organization, city-state and empire, is remarkable. In the long run neither proved viable. The city-states had disappeared as military and political actors by the end of the first millennium B.C., and though the institution reappeared in the Middle Ages, in the Mediterranean world the revival was limited to Italy. The Roman Empire, in which the traditional republican form of government soon gave way to imperial autocracy, remained united until near the end of the fourth century A.D., but was then divided; the western part fell to invaders from the north in the fifth century, whereas the eastern part survived in some fashion until 1453, though losing increasing amounts of territory to Muslim invasion. Since the fourth century the Mediterranean has never again been united under a single state, Roman or other.

Mediterranean Culture

It was against this shifting political background that the cultural history of the Mediterranean world unfolded. The Mediterranean in 1000 B.C. must have been a world of many cultures, though none of them outside Egypt and the Near East could be called civilizations. It would hardly have seemed likely at the time that the most important of these cultures in historical terms would prove to be that of the Greeks. The Mycenaean civilization of the Bronze Age had collapsed a couple of centuries before, and the Iron Age had yet to generate a new one; there is no evidence that the Greeks in this period possessed any form of writing. But two things can perhaps be identified that gave their culture a degree of advantage. One was the relative ethnic and linguistic homogeneity of Greece, and the other was the beginnings of the Greek colonization overseas. These features meant that Greek culture would have a larger constituency and a wider distribution than any likely competitors.

The literary culture that developed among the Greeks in the mid-first millennium B.C. shared a couple of significant features with those of contemporary India and China. In all three the Bronze Age past played a central role in the culture—the Greek equivalents of the Vedas and the older Chinese classics being the Homeric epics. But alongside this archaic heritage, in all three cultures a novel interest appeared around the middle of the millennium: rigorous philosophical and scientific thought. One feature of Greek culture, however, set it apart from Indian, Chinese, and other literary cultures of the day: its close relationship to the city-state. Greek society lacked the kind of powerful bureaucracy that elsewhere might provide a locus for the formation of an elite culture (for example, in ancient Egypt); its temples were not prominent in the appropriation of economic resources (as was the case in the ancient Near East), and there was no entrenched and hereditary priesthood (as there was in ancient India). Greek culture was thus marked by a focus on the political that is largely absent from Indian culture as it has come down to us; renouncing the world became a prominent theme among the Greeks only with the demise of the independent city-state, and even then a philosophy of renunciation like Epicureanism never became a mass movement in the manner of Buddhism. At the same time this focus was significantly different from that of political reflection in ancient China, where even under conditions of fragmentation the scale of political organization was much larger. In this way the culture of Greece stood out in being what we might call a citizen culture.

The spread of Greek culture to non-Greeks was a prominent theme in the history of the ancient Mediterranean world, and we can look at it in terms of a distinction we have used before. One major process by which the culture spread was conquest effected by the Greeks themselves or, more, precisely by their Macedonian overlords. In 334 B.C. the Macedonian ruler Alexander set out to conquer the Persian Empire. When he died in Babylon, in 323, he had brought Macedonian rule, and with it Greek cultural dominance, to a region extending from Anatolia and Egypt in the west to Central Asia and northwest India in the east; and over much of this territory he had established Greek urban settlements. Greek elite culture was accordingly to have a long history in this region, but over the centuries it was gradually to lose its hold. Only in Anatolia did the mass of the population become Greek-speaking, and even this gain was eventually to be reversed.

In the end the adoption of Greek culture by politically independent non-Greek societies was to prove of much greater historical importance. The rulers of Lydia seem to have been honorary members of the Greek cultural world, and there is evidence both here and around the Mediterranean for the adoption of elements of Greek culture by indigenous elites in the neighborhood of Greek colonies; thus the Gauls of southern France are said to have learned from the Greeks "a more civilized way of life," tilling their fields, walling their towns, living by law rather than force, and cultivating the vine and olive. A particularly thoroughgoing example of such assimilation took place in the first half of the fourth century B.C. when Mausolus, who ruled the Carians in southwestern Anatolia, set about imposing the Greek way of life on his people, among other things by forcibly resettling them in cities. But the crucial reception of Greek culture took place in Italy. As we have seen, the Etruscans acquired the alphabet from the Greeks and passed it on to the Romans. But this early borrowing, though far from isolated, does not seem to have been part of a wholesale adoption of Greek literary culture, and led rather to the definitive establishment of Latin as the local literary language. It was at a later stage that the Romans came into much closer contact with the Greek world through their conquest of Greece and the eastern Mediterranean. The Roman elite of the second century B.C. then embarked on a massive assimilation of the civilization it had conquered. As a result, any educated Roman knew Greek, and the Greek heritage became an integral part of Roman culture. Yet the primary literary language of the Romans was still their native Latin. It was this Greek culture in Latin dress that was spread by virtue of Roman conquest throughout the western Mediterranean.

What is at first sight surprising in all this is that the lands of most ancient civilization, Egypt and the Near East, should not have played a more prominent role in the westward spread of civilization in the first millennium B.C. Mesopotamia, of course, was some way away from the Mediterranean, and Egypt, though located on its southeastern shore, had never shown much interest in maritime expansion; it lacked the timber resources to build ships, and in any case its civilization never spread far beyond the frontiers of Egypt. It is more puzzling that the Phoenicians did not play a greater role in the outcome, as opposed to the initiation, of the process. But their expansion clearly lacked the demographic pressure that fueled that of the Greeks; and their one major colony, Carthage, was defeated in its struggle with Rome. The result was that what the ancient Mediterranean bequeathed to Europe was Greek culture in its Latin form.

But this was not quite the end of the story. As we saw in chapter 7, the religion to which the Roman Empire was to succumb in the fourth century A.D. was ultimately the legacy of a people whose language was virtually identical to that of the Phoenicians.

Christianity is a fine example of the way in which a religion can snatch victory from the jaws of defeat by taking its greatest liability and making a feature of it. The founding figure, Jesus of Nazareth, was a Jewish popular preacher and miracle worker. He told his followers that the kingdom of heaven was at hand

(Matthew 4:17) and urged them to adopt a touchingly altruistic morality (Luke 6:27–35). For largely political reasons he was subjected to the cruel, but not unusual, Roman punishment of crucifixion. This meant that he died like a common criminal alongside a couple of robbers, unable to save himself—so bystanders taunted—let alone others (Matthew 27:42). Yet this disaster proved to be the beginning rather than the end of a new religion. The followers of Jesus claimed that despite appearances he had been the Son of God, or God incarnate, and that he had willingly died on the cross to redeem mankind. Most of those who initially responded to this message were naturally local Jews, but the existence of a widely scattered Jewish diaspora meant that word spread quickly all over the Roman Empire. At the same time the early church was astute enough to customize the new faith for Gentiles by allowing them to join without taking upon themselves the irksome burdens of the traditional Jewish religious law (Acts 15:19–20). The outcome was that gradually the movement became overwhelmingly Gentile.

Like the Buddhists, the early Christians argued among themselves, and in due course they split up into a number of irreconcilable sects, despite (or because of) the deliberations of a series of councils. But in organizational terms the Christians differed from the Buddhists in significant ways. Whereas monks made up the core of the Buddhist community from the outset, Christian monasticism did not appear until some three centuries after the death of the founder. It then became a standard feature of Christian life, albeit eventually rejected by the Protestants; but it never took over. Instead, the core of the church was and remained a set of lay congregations under the authority of their priests, who in turn came to be subject to that of the local bishop. A further contrast with the Buddhists was the extent of the overarching hierarchy that developed in Christianity above this local level. Buddhist sects in the early centuries had some kind of succession of patriarchs; but these rather shadowy figures do not seem to have matched the power of the handful of Christian patriarchs who came to hold sway over the church from their seats in the major cities of the empire. Of these by far the most successful were the popes, who from their base in Rome were able to consolidate their authority over the western half of the empire. They thereby brought into being the Catholic church. This remarkable institution, despite its heavy penetration by the interests of rulers and others, has proved the most impressive and durable nongovernmental organization in the history of the world.

In the meantime the church's relations with the Roman emperors and their successors had changed drastically. The Christians had never plotted to overthrow the pagan empire; they were not like the Jewish Zealots, whose monotheist intransigence fed into a wildly imprudent rebellion against Roman rule in A.D. 66–70. But the Christians nevertheless picked a quarrel with the Roman state: as part of their monotheist rejection of other gods, they refused to participate like loyal subjects in the standard rituals of emperor worship. Such attitudes led to intermittent persecutions of a kind that the Buddhists were usually spared, but not to a terminal catastrophe. Thus, by the early fourth century, there were apparently enough Christians for it to make sense for Constantine, a claimant to the throne locked in conflict with his rivals, to identify himself with the Christian cause. He won his civil war, and by the end of the century Christianity was firmly established as the state religion.

II. The Background to Athenian Democracy

Most of the works of Greek literature that have reached us at all have done so through continuous literary transmission since the time when they were written. But there is another route through which ancient works occasionally surface: the Egyptian desert. For example, four rolls of papyrus came to light there in the late nineteenth century, and turned out to contain the bulk of a short work by Aristotle (d. 322 B.C.) on the Athenian constitution. At one point this archetypal academic summed up the history of the constitution with a rather dry listing of the eleven occasions on which it had been changed—the kind of thing it does students good to learn by heart. The first of these changes was a slight deviation from absolute monarchy in the reign of Theseus, a more or less legendary figure of the Bronze Age. The rest were crowded between 621 B.C., when Draco drew up his notorious ("Draconian") law code, and 403 B.C., when democracy was restored after a decade of mayhem in the generation before Aristotle was born; here the events are historical, though the earlier dates are a bit arbitrary. Why, then, did the Athenians change their constitution ten times in this period? And why, for that matter, did they have a constitution at all?

The best place to start looking for an answer is the crisis of 596 B.C. By this time dynastic kingship in the style of the Bronze Age was a thing of the past. There was still a "king" in Athens, but he was just one of a trio of magistrates who held office for a decade, later only for a year; the other two were a military commander and an official with a wide range of duties, the archon. The magistrates were chosen for their birth and wealth, and after they had served their terms they joined the Council of the Areopagus, a powerful body the formal task of which was to watch over the laws. Thus, unlike the Romans, the Athenians had not expelled their kings, but they had long ago cut them down to size. Athens at the beginning of the sixth century was what we, using a Roman term, would call a republic.

The problem in 596 was not royal despotism but the discontent of the common people. This is interesting. In our survey of Old World civilizations so far, we have taken the common people for granted as an indispensable part of the infrastructure of elite culture. Without peasants toiling away, how would Near Eastern kings have built their palaces, Brahmins cultivated their Vedas, or Chou aristocrats accumulated that enormous tonnage of bronze vessels?

Yet at no point have we been forced to ask how the mass of the population might have felt about its role in all this; and if we were now to try to make up for this omission, we would not find much evidence to help us—though in China, once we reach the imperial period, we encounter occasional cataclysmic outbreaks of peasant rebellion. In Athens, by contrast, the common people were a force to be reckoned with on a continuing basis, and in 596 they had a lot to be discontented about.

At this time the constitution of Athens was, as Aristotle puts it, an extreme form of oligarchy. The land was owned by the few, and the masses were their tenants; if their rent fell into arrears, they were liable to be enslaved, together with their children. Such conditions were more than ordinary people were prepared to put up with in Athens, and the result was prolonged civil strife. It is clear from this that the Athenian elite could not hope to suppress the rebelliousness of the masses once and for all, nor could they simply wait for the storm to pass; some drastic adjustment was needed. So by agreement between the two sides a certain Solon, known for his wisdom, was made archon with the power to reform the constitution. With regard to the agrarian problem, his compromise was to cancel debts and ban the enslavement of debtors, but to leave the land in the hands of the rich and powerful so that, as he put it, they should "suffer no unseemly shame." At the same time he issued a new law code to supersede that of Draco, liberalized the rules under which magistrates were selected, and instituted a new council of four hundred; the poorest class were allowed membership of the popular assembly and a role in the lawcourts. Having done all this, Solon left the city for ten years of self-imposed exile. The result was turmoil.

The man who delivered the Athenians from this predicament was the tyrant Peisistratus. A tyrant in a Greek city-state was someone who seized power and established his personal rule. Peisistratus first did so in 560, and the regime he eventually established lasted till 510. But the manner in which he exercised his power was as instructive as the preceding disorder. One of the foundations of his success was sword control; he disarmed the people by a trick and told them to go home and mind their own business while he attended to affairs of state. The other was his practice of lending money to the poor so that they would leave the city and occupy themselves with agriculture (he had made a fortune in a part of Thrace known for its gold mines). These measures, together with his attractive personality, scrupulous observance of the laws, and considerable political skills, won him the support of most notables and the majority of the common people. The key to his success was thus getting the common people out of politics while making them economically contented—a bargain many of us might feel inclined to accept.

Peisistratus' sons lacked their father's personal qualities, and with Spartan help the Athenians eventually got rid of the family. Elite politics now resumed, with two factions and two leaders. Cleisthenes, the leader who happened to be losing, decided to play the popular card, offering the masses democracy in return for their support. His oligarchic antagonist responded by playing the Spartan card, but the Spartan intervention went awry: the populace flew to arms—somewhere along the way they had clearly recovered them—and the Spartans, after a couple of days walled up in the Acropolis, agreed to leave. The people had won, and in 508 B.C. Cleisthenes put through an elaborate reform of the constitution. With some later fine-tuning, the democratic order thus instituted lasted for almost a century, until in 413 it was destabilized by a major Athenian defeat in the course of the war with Sparta. This led to unstable oligarchic regimes and another disarming of the people. After the war was over, the democracy was restored in 403—the last change noted by Aristotle—and it was still flourishing decades later when the Macedonians took over. The central institution of this democratic order was the sovereign assembly of the citizens of Athens.

In the turbulent politics of the Greek city-states, democracies were not usually any more stable than other regimes. Why, then, was that of Athens so successful? No doubt part of the explanation was a well-designed constitution, spelling out the interlocking roles of the magistrates, the council, the assembly, and the lawcourts. But the underlying reason why this constitution had a chance of working over the long run was very simple: for a Greek city-state, Athens was about to become unusually rich. In 483 the city started to derive great wealth from the silver mines located on its territory, and in 478 it began to receive tribute payments from the members of the league of Greek city-states it had established against the Persians—less politely, the Athenian empire. (In fact, the two sources of income were linked: the firstfruits of the silver mines had prudently been reinvested in the naval power on which the Athenian empire was to depend.) With resources such as these, Athens could keep a large number of potentially discontented citizens happy without turning on the rich, and the conflict between elite and masses could thus be kept within bounds. Aristotle underlines the point in his usual professorial style: "Examine the history of other states, and you will find that when a democracy comes to power, so far from delving into its own pocket, it nearly always makes a general redistribution of the land."

What is the moral of all this? First, a theme that recurs throughout the story is the peculiar vulnerability of the city-state to the socioeconomic grievances of its masses. For an ambitious politician, feeling their pain might be an avenue to political power, and failure to feel it could be disastrous. This did not, of course, mean feeling everybody's pain. Athenian wealth and power led to the presence of large numbers of slaves and resident aliens, who played no part in the city's politics; nor did women. But even when we set these groups aside, there was still a populace to be reckoned with. It was not like this in the monarchies of the Near East, India, China, or the later Mediterranean world. Here, then, is one reason for the distinctive focus of Greek (and Roman) political thought.

Second, the effect of this vulnerability, under normal circumstances, was to make city-states pretty much ungovernable. Only under special conditions do we encounter significant exceptions. In Sparta the citizen body as a whole exploited a large rural population known as the helots; so it was the helots, not the Spartan masses, who repeatedly rose in revolt. In Carthage the profits of the city's maritime activities in the western Mediterranean must have provided abundant resources; Aristotle, who had a high opinion of the Carthaginian constitution, noted approvingly that it retained the loyalty of the masses. And yet even Rome, awash with the loot of imperial expansion, went through periods of acute social tension. In the world of city-states at large, democracy seems to have been even less of a route to stability than monarchy or oligarchy. Under such conditions it took a remarkable commitment to political freedom to prefer the turmoil of democracy to the benign tyranny of a Peisistratus.

III. Attic Black- and Red-Figure

Pottery, as we know from other contexts, is one of the elements of material culture that survives best. Figure 18 shows four Attic pots; Attica was the corner of Greece attached to Athens, so that in effect these pots are Athenian. They span some six centuries in a highly discontinuous way. The oldest (at the far left) and the most recent (at the far right) are perfectly decent pots, and by no means plain. But they do not give us much insight, other than technical, into the culture that produced them. For all most of us would know, they could just as well have come from Ecuador or Cambodia. In between, a remarkable window onto Athenian culture opens with the coming of black-figure and closes with the passing of red-figure. For some decades the two styles were produced concurrently, but for most purposes red-figure drove out black.

Fig. 18: Four Attic pots. ***(Left):*** **a Geometric oenochoe, or wine jar (about 750 B.C.);** ***(right):*****a black-figure amphora (about 530 B.C.).**

The difference between black- and red-figure was simple: in black-figure the pigmentation was applied to the figures and the background left plain, whereas in red-figure it was the other way round. The technical requirements for manufacturing them were not so simple. All potters need two things if they are to produce superior wares: good clay and the ability to fire it at high temperatures. Attic clay was of good quality, but what made it special was the amount of iron it contained. The chemistry of iron is such that it forms oxides of differing colors. The distinctive feature of the firing process for black- and red-figure was that the potters had to manipulate this chemistry by combining control of temperature with switches between oxidizing and reducing atmospheres. If they succeeded, the happy result was to leave the right oxide on the right part of the pot. That they did not always succeed is apparent from the misfired pots that survive.

Black- and red-figure are a boon for even the most philistine

***(Left):* a red-figure bell crater, or mixing bowl (about 470 B.C.); *(right):* a West Slope amphora (first half of the second century B.C.).**

archaeologist. Such pronounced forms of decoration lend themselves to frequent changes of style, one might almost say of fashion. The result is that a black- or red-figure pot can be dated with unusual precision. This is why in figure 18 the datings of the pots in the middle leave much less leeway than is the case with the ones on either side. Recently an ancient burial ground in Athens yielded the cremated remains of some two hundred young men; the pottery found with them (including red-figure vases) made it possible to date their deaths to the 420s, the first decade of the long war between Athens and Sparta. In many parts of the world, by contrast, pottery may not vary much from one century to the next.

A more lively aspect of this form of ceramic decoration is its capacity to show us ancient scenes that would otherwise be denied to us. The red-figure detail in figure 19, for example, shows what a loom looked like in ancient Greece (unlike the kind that craft-conscious people are familiar with today, it is vertical: the warp is suspended from the top of the loom). As often in vase painting the

Fig. 19: Penelope at her loom.
Attic red-figure, about 440 B.C.

scene is set in the Bronze Age, the woman at the loom being Odysseus' wife, Penelope. But the objects depicted are likely to be contemporary with the painter.

More interestingly, the scenes on the pots sometimes show us what the pots themselves were for. In ancient Athens, as in many societies, knowing how to get drunk in good company was an art no elite male could afford to be without. Black- or red-figure pots had key parts to play in this. Like Chinese bronze vessels, they came in sets. There had to be a decanter—called an amphora—for the wine, a jar for the water, a bowl in which to mix the wine and water, a jug with which to serve the mixture, and wine cups for the individual drinkers. The mixing needs a word of explanation: the ancient Greeks cut their wine with water, just as their descendants still did in the last century. As to the cups, they tended to be broad and shallow (as in figure 20, above); one is tempted to imagine that the attempts of the drinkers to hold them upright gave rise to increasing amusement as the evening progressed.

Naturally women of good family would not be present at such parties, but courtesans might be. Figure 21 (on the left) shows one wearing nothing but a necklace and a band around one of her thighs; contrast her shameless gaze with the bashfully downcast eyes of a respectable woman bidding farewell to a warrior (on the right). But the services courtesans supplied on such occasions were not confined to sexual companionship. Figure 20 (below) shows a young man who has obviously drunk too much and is throwing up; a sympathetic courtesan holds his head, though prudently keeping a certain distance (she only works there).

If today you visit a museum with a significant collection of Greek antiquities, pots are a large part of what you see. The reason for this is simple. Apart from marble—whether used in sculpture or in architecture—clay is the only medium of fine art that has survived in quantity. Indeed, in one way it has proved a much better survivor than marble: the later inhabitants of Greece assiduously recycled ancient marble to make lime, but had no comparable use for ancient pottery. The result has been to give us a rather misleading impression of the status of vase painting in its own time. Painting on wooden panels and walls was a serious art in ancient Greece, and there must have been many fine examples of it in Athens; in this company vase painting was something of a poor relation. Moreover, the clay pots themselves may to an extent have been poor relations of bronze vessels used by the cream of the elite. With rare exceptions these too have perished; bronze, even more than marble, invites recycling, and for the Greeks, as for most peoples other than the Chinese, it was too valuable a material to be buried with the dead on a large scale. Thus Greek ancestors had for the most part to make do with pottery, which is why we have so many intact Greek pots today.

Fig. 20: ***Above:*** **an Attic red-figure wine cup (about 490–480 B.C.).** ***Below:*** **looking down on the scene on the inside of the cup.**

Fig. 21: *Left:* a sherd from an Attic red-figure wine cup showing a courtesan with a young man (about 500–490 B.C.). *Right:* a sherd from an Attic red-figure pot showing a woman bidding farewell to a warrior (about 430 B.C.).

The craftsmen who painted these vases certainly had their pride. Some signed their pots, and one even included negative advertising directed against a rival. At the same time fine vases were widely appreciated; thus in the fifth century B.C. Attic pottery was exported all over the Greek world, and even beyond it. But Greek literary culture paid little attention to pottery, and were it not for archaeology we would know nothing of Attic black- and red-figure. Nor did any medieval antiquarian ever wax enthusiastic over it. Only in the seventeenth and eighteenth centuries did a serious interest in Greek vases begin to develop in western Europe, and even then it took some time to dispel the notion that they were Etruscan. The bottom line is that we have been looking at an aspect of Greek culture that matters more to us than it did to them.

CHAPTER II

WESTERN EUROPE

1. An Unlikely Corner of the World?

Western Europe, like India, is a peninsula of Asia, but geography does not really tell us where the peninsula begins. For our purposes a maximum definition would include all the territory that came to form part of western Christendom in the Middle Ages (minus the transient fruits of its adventures in the eastern Mediterranean). This domain stretches from Spain and Italy in the south to Scandinavia in the north, from Ireland in the west to Lithuania in the east. But Scandinavia and eastern Europe will be of relatively little concern to us, and the Mediterranean south is already familiar. Hence most of our attention in this chapter will be directed to northwestern Europe; this is a region we have not yet encountered, and one that will play a key role in a later chapter. But other parts of Europe will also make their appearances, particularly the south.

We are used to thinking of Britain, the Low Countries, and northern Germany as among the most densely populated parts of the world. In geographical terms this is rather startling. We are talking about a region that lies at the same latitude as Labrador and the Kamchatka, two forlorn and frigid peninsulas that typify a vast swath of territory located too close to the Arctic for human comfort. North America and Asia in general are at best thinly populated at this latitude; there hardly seems a place in such company for "England's green and pleasant land." Yet Europe, and especially northwestern Europe, stands out in sharp contrast to North America and Asia. Unlike Labrador and the Kamchatka, this region enjoys a temperate climate: a winter that is not so cold as to preclude the survival of a dense population, a summer that allows an adequate growing season, and plenty of rain. This climate is the gift of the Atlantic or, more precisely, of the way seawater circulates in it, bringing warm tropical water to the shores of Europe. Like so much else in the world, this system is likely to be contingent. There may have been no such pattern of circulation in the last ice age, and there may be none when the climate takes its next lurch. Perhaps nowhere is the idea of a fleeting Holocene window of opportunity more in place than in northwestern Europe.

Within the window, however, this part of the world possessed several geographical advantages. Like the northern Mediterranean, it had a highly indented shoreline that encouraged navigation; like China, it had good navigable rivers. North of the Alps and the Pyrenees it was clear of massive mountain ranges; the hills and plateaus of France and Germany, and still more those of the far northwest, were a great deal older than the Alps and thus heavily eroded (see map 9A). Plains were in generous supply. Those that had been covered by glaciers in the last ice age tended to have poorer soils, but they lay mainly to the northeast; much of the northwest was free of this, and its soils included considerable areas of loess, just as in the Yellow River valley of China. At the same time the region did not suffer the disadvantage of being remote from the Near East. It was less distant from it than China, and unlike sub-Saharan Africa it was linked to it by the Mediterranean, not separated from it by desert.

ICELAND
Norwegian Sea
Norsemen
(Vikings
North Sea
SCOTLAND
IRELAND
BRITAIN
ENGLAND
WALES
ATLANTIC
OCEAN
Clacton
Canterbury
KENT
LOW COUNTRIES
HOLLAND
Dutch
Flemings
GERMANY
Aachen
BELGIUM
NORMANDY
BRITTANY
Franks
Paris
0
200 miles
0
200 kilometers
FRANCE
(GAUL)
Bordeaux
ALPS
Basques
PYRENEES
PORTUGAL
SPAIN
ARAGON
IBERIAN
PENINSULA
Mediterranean
Sea
Rome
ITALY

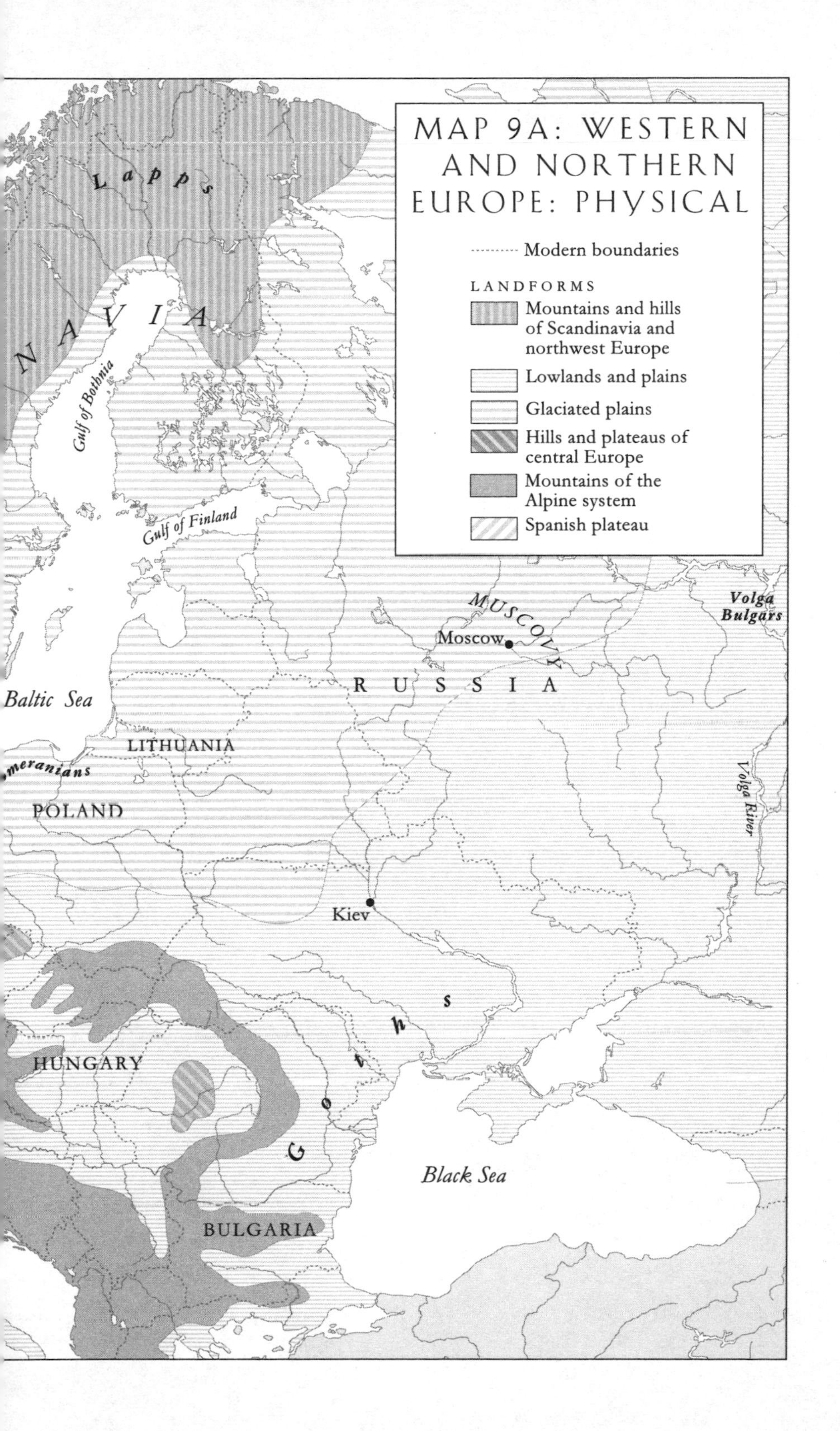
MAP 9A: WESTERN AND NORTHERN EUROPE: PHYSICAL
Modern boundaries
LANDFORMS
Mountains and hills of Scandinavia and northwest Europe
Lowlands and plains
Glaciated plains
Hills and plateaus of central Europe
Mountains of the Alpine system
Spanish plateau
Lapps
NAVIA
Gulf of Bothnia
Gulf of Finland
Baltic Sea
LITHUANIA
POLAND
MUSCOVY
Moscow
RUSSIA
Volga Bulgars
Volga River
Kiev
Goths
HUNGARY
BULGARIA
Black Sea

ICELAND
Norwegian Sea
0 200 miles
0 200 kilometers
ATLANTIC OCEAN
SCOTLAND
North Sea
IRELAND
BRITAIN
ENGLAND
WALES
Norsemen (Vikings)
LOW COUNTRIES
HOLLAND
Dutch
Clacton
Canterbury
KENT
Flemings
BELGIUM
Aachen
GERMANY
NORMANDY
BRITTANY
Paris
Franks
FRANCE (GAUL)
Bordeaux
ALPS
Basques
PYRENEES
SPAIN
ARAGON
Mediterranean Sea
IBERIAN PENINSULA
Rome
ITALY

Voguls
Lapps
V I A
Gulf of Bothnia
Gulf of Finland
MUSCOVY
Moscow
Volga Bulgars
R U S S I A
Volga River
LITHUANIA
OLAND
Kiev
Goths
IGARY
Black Sea
BULGARIA
MAP 9B: WESTERN AND NORTHERN EUROPE: NATURAL VEGETATION
Modern boundaries
VEGETATION
Mountain and tundra
Northern forest (taiga)
Temperate forest
Mediterranean vegetation
Steppe
Heathland

Northwestern Europe as a Laggard

Yet, in the history of civilization, northwestern Europe appears as a laggard. It never developed a civilization of its own; nothing that emerged on the northern plains in the course of the Bronze Age could compare with Shang China or the even older civilization of the Indus Valley. Nor did northwestern Europe import an alien civilization of its own free will, as did Southeast Asia. There were, of course, contacts with the more advanced cultures that appeared in the western Mediterranean in the first millennium B.C., but the Hellenization of the Gauls of southern France does not seem to have extended significantly into the hinterland. When the Romans conquered Gaul in the first century B.C., they conquered tribes, not city-states; and like the Britons in the next century, the Gauls then had civilization thrust upon them. Farther to the east were the Germans. They showed the same lack of interest in adopting the civilization of the Romans, and the Romans had better things to do than to force it on them. What joy could there have been in attempting to conquer a land of "forbidding landscapes and unpleasant climate," as the Roman author Tacitus described Germany in the first century A.D., "thankless to till and dismal to behold"? Beauty, to be sure, is in the eye of the beholder, but we will come back to the question why the forests of Germany—and not only Germany (see map 9B)—were thankless to till.

In its earlier prehistory northwestern Europe was by no means such a laggard. Its record in the Upper Palaeolithic was quite as impressive as that of any other part of the world, and not just because it has been better studied. The region did not invent farming or metalworking for itself, but its adoption of these techniques was rapid by the standards of the time. One example is the spread of farming over the loess soils from Hungary to Belgium within a period of a couple of centuries in the sixth millennium B.C.; much as in the Near East, we have no evidence of hunter-gatherers in Europe in historical times. Another example is the arrival of iron metallurgy in Britain by the fifth century B.C., only two or three centuries after it had become established in Italy. Indeed, as prelit-erate cultures go, those of northwestern Europe were marked by considerable sophistication. But civilizations they were not.

The Roman conquest was thus a major event for the part of northwestern Europe that was subjected to it. We can still see its impact in the language map of the region today. This map is made up of features of varying antiquity. By far the oldest is likely to be the existence of Basque, now confined to a small territory in southwestern France and northern Spain, but once considerably more widespread. It is the only surviving non-Indo-European language of western Europe; it may even have been more or less where it now is since Upper Palaeolithic times. The next-oldest stratum is the Celtic languages, a branch of Indo-European of much later vintage. With the Celtic presence we come to the first period in the past of northern Europe for which we can begin to match archaeological evidence with the record of historical sources—though as yet these sources derive only from the Mediterranean world and are very patchy. In this way we know that in the centuries around the middle of the first millennium B.C., the Celts came to occupy a vast area of northern Europe and also intruded into each of the four Mediterranean peninsulas—violently in the three cases where we have direct historical testimony, and presumably also in Spain. Hence Celtic languages were well established in Gaul and the British Isles at the time of the Roman conquest. Yet in France, just as in Italy and Spain, and very likely in southeastern Britain, the effect of the conquest was to spread Latin at the expense of the local languages; initially this must have affected mainly the urban elite, but eventually Latin displaced Celtic even among the peasantry. Today all that is left of the once massive dominance of Celtic in northwestern Europe is the "Celtic fringe" of Brittany and the British Isles—the rest of the region speaks languages derived from Latin or Germanic. So it is clear that Roman conquest broke the cultural continuity of Celtic Europe; only in peripheral regions beyond the imperial frontiers—above all, in Ireland—did an integral Celtic culture survive.

A major change that affected the region toward the end of the Roman period was the spread of Christianity. As elsewhere in the empire, it was the adoption of the new religion by the emperor Constantine in the first half of the fourth century that initiated a general conversion. By the time the Roman Empire in the west came to an end in the fifth century, the Christian church was a major institutional and intellectual force in Gaul. The establishment of the church was crucial for the future of European culture. It gave canonical status to a heritage that would otherwise have been utterly irrelevant to the peoples of the region, that of ancient Israel. But it also conserved the literary heritage of the Greco-Roman world. Christianity no doubt lowered the tone of elite life in late antiquity; but thanks to its bishops and monasteries, its organizational resilience, and its commitment to the survival of literate culture, it may have prevented the kind of discontinuity that followed the collapse of the civilizations of the Indus Valley or Mycenaean Greece. Without the church, the best we could imagine might have been the slow accretion of a new literary heritage around Germanic runes and Irish ogham—folk scripts that the early Germans and Irish somehow derived from the writing systems of the Mediterranean world.

The church apart, the civilization established by the Roman conquest of northwestern Europe was not particularly robust. The Romans certainly left traces wherever they went. The main roads of Anglo-Saxon England, such as they were, originated as Roman roads. But Rome was a Mediterranean power, and in northwestern Europe the Roman presence, for all its linguistic impact, did not have the same cultural density as it did in the south. And in any case, it embraced only a limited territory; it did not extend to Germany, let alone Scandinavia or eastern Europe.

It was this thinly civilized and severely circumscribed territory that was overrun, together with the western half of the Mediterranean world, by a variety of Germanic peoples in the fifth century A.D. In Italy, Spain, and North Africa, Germanic conquerors established kingdoms that lasted a couple of centuries, more or less; they left the language, ethnic identity, and cultural traditions of the subject Roman populations somewhat the worse for wear, but otherwise pretty much unchanged. In Britain, by contrast, the impact was drastic, even though the sea might have been expected to provide some protection against heavy Germanic settlement. Outside the western fringes of the island Germanic replaced Latin or Celtic, and Germanic ethnic identities prevailed; south of Scotland and east of Wales the land was divided among half a dozen petty kingdoms, not to be united into the country we know as England till the ninth century. In the course of this transformation, Roman culture and Christianity had been effectively erased, and both had to be reintroduced by Christian missionaries from the Continent at the end of the sixth century. Gaul—or as it now became, France—presents an interestingly intermediate case. Latin prevailed over the Germanic language of the conquerors, though in the process of becoming French it was heavily influenced by Germanic. In ethnic terms, however, it was the Frankish conquerors who prevailed: the inhabitants of France eventually came to be known as Franks or, as we now say, French—not as Romans, still less Gauls. Once the Frankish king and his people had converted to Christianity in 493, the place of what remained of Roman culture in its Christianized form was reasonably secure; but there was not an enormous amount of it in northern France, where the center of Frankish power was located.

For nearly three centuries the Frankish kingdom was ruled in some fashion by the Merovingian dynasty (the longhaired kings, a nice contrast to the skin head kings who were their contemporaries among the Bulgars). But in 751 this dynasty was replaced by a family that already held effective power in the kingdom, the Carolingians. The long reign of the most successful Carolingian ruler,

Charlemagne (768–814), is one of unmistakable significance in the history of northwestern Europe. For the first time since the fall of the western Roman Empire, we sense the emergence of a new order in the region. The Frankish kingdom, already a sizable state, was now further enlarged to the east and south. Great efforts and considerable patronage were expended on the revival of Latin culture. And in a fine symbolic moment in 800, Charlemagne on a visit to Rome was crowned Roman emperor by the pope. Yet there was no attempt to restore any geopolitical semblance of a Roman Empire around the Mediterranean, a sea that was now largely dominated by the Muslims. Instead, Charlemagne ruled from Aachen, a town so far to the north that it forms part of the latitudinal band we compared to Labrador and the Kamchatka. This is perhaps the first intimation that it was even possible for such a region to be at the center of a civilization, and not just on its remote northern periphery.

Charlemagne's new order was, however, of short duration. The unity of the Frankish state did not long survive his death, and political conditions in northwestern Europe became increasingly chaotic in the course of the ninth and tenth centuries. In addition to internal disarray the region was subject to several external threats, and proved signally ineffective in parrying them. To the south were the Muslims. They had already conquered the bulk of Spain in the early eighth century, and went on to help themselves to most of the islands of the western Mediterranean. They did not threaten large-scale territorial conquest of Frankish lands, but for the best part of a century a group of Arab marauders with a base in the south of France wrought havoc in the surrounding territory. To the north were the Scandinavians—the Vikings, or Norsemen—whose maritime raiding terrorized the coasts of the mainland and the British Isles. In some regions they settled in significant numbers; in France they gave their name to Normandy, where they contributed to the formation of a particularly effective Frankish-style military aristocracy. To the east were the Hungarians, nomadic invaders from the steppes who arrived in the territory that was to become Hungary in the later ninth century, and raided widely for some time thereafter. Under these conditions the grand cultural pretensions of Charlemagne's reign were unsustainable.

This takes us more or less to A.D. 1000. How should we sum up the ground we have covered to this point? Like Southeast Asia, northwestern Europe had not developed a civilization of its own, and instead had acquired one from elsewhere—except that while the Southeast Asians voluntarily adopted theirs, the northwestern Europeans had it imposed on them. Just as in the case of early Southeast Asia, there is nothing in the history of northwestern Europe up to this time to suggest that it was destined to play a major role in world history. We reluctantly dispatched Southeast Asia in a paragraph while looking at India, and it is hard at this point to see why we should not have treated northwestern Europe in the same way, as an appendage of the Mediterranean world. Only the location of Charlemagne's capital in the far north suggests otherwise, and that could have been as much a passing anomaly as a sign of things to come.

Northwestern Europe Catches Up

If the first millennium was a rather uninspiring period in the history of northwestern Europe, why should the second have been any different? A significant part of the answer lies in a poorly documented development of the early Middle Ages that had enormous implications for the future. This takes us back to Tacitus' dismissal of Germany as "thankless to till." The farming package adopted in northern Europe in Neolithic times had been the product of a Mediterranean climate. In the rather different conditions of the north, it is perhaps remarkable that it worked as well as it did (contrast the need to domesticate new crops locally in sub-Saharan Africa). The primary limitation of this agricultural heritage was in fact more technological than biological: the simple scratch plow that was adequate for the dry, light soils of Mediterranean Europe was ineffective when pitted against the heavier soils of the northern plains and valleys. Early agriculture in the north had thus tended to be concentrated on land that today would be regarded as marginal, and to leave the richest soils uncultivated. The development of a heavier plow suitable for northern conditions seems to have taken place around the sixth century A.D. As the new plow became widespread in the centuries that followed, and peasant society reshaped itself around it, the productivity of agriculture in the north greatly increased. The effects of this were not, of course, confined to the peasantry. The new capacity of northwestern Europe to support urban populations and elites made possible a society that was in several ways very different from its predecessor.

Perhaps the most obvious of these changes was the much greater military efficacy of northwestern European society from the eleventh century onward, as feeble defense gave way to formidable aggression. This was most evident on the frontiers of Christendom. Knights from the northwest played roles of varying prominence in the reconquest of Spain and Sicily from the Muslims, the establishment of Crusader states in the eastern Mediterranean, and the conquest of pagan territory in the northeast. But the military edge of the old northern cradle of the Frankish state was also manifested within the boundaries of the Christian world by conquests in England, Ireland, and the south of France. The military basis of these developments was a particular use of an invention of the early Middle Ages that reached Europe from the east—the stirrup. The Franks and those who fought like them used it to perfect a style of warfare in which the ultimate weapon was the disciplined charge of heavily armored cavalry. As a Byzantine princess remarked of the Franks, "a mounted knight is irresistible; he would bore his way through the walls of Babylon." An army of this kind went well with the domination of peasant society by a warrior aristocracy.

Another significant trend in this period was the emergence or consolidation of more powerful and centralized states. Perhaps the most obvious case was that of England, where the process was accelerated by the establishment of a kingdom of continental origin through the Norman conquest of 1066. Within the British Isles this kingdom went on to conquer Wales and most of Ireland, though not Scotland, which succeeded in retaining its independence through a process of defensive Normanization. On the Continent the English state long retained and sometimes greatly expanded its territories. But it finally lost them in the mid-fifteenth century, and what emerged from the Middle Ages was a strong but insular monarchy. The rise and consolidation of the French state was slower and somewhat unsteady—what was good for England was bad for France—but in the end it produced a stable domination of a large and populous territory. Both these kingdoms were in some sense nation-states by early modern times. The Spain that emerged from the demise of Muslim power in the Iberian Peninsula tended in the same direction, but the formal unification of the entire peninsula lasted only a few decades; even then the Portuguese kingdom was never fully absorbed, and that of Aragon only in the eighteenth century. In Germany and the Low Countries the eventual failure of rulers to establish solid and substantial kingdoms led to much greater political disunity, and Italy likewise remained fragmented. This did not, however, render these regions unimportant: they included the two most urbanized societies in Europe, those of northern Italy and the Low Countries.

But the most interesting feature of northwestern European states was not their size; rather it was the character of their relations with their subjects. At least two themes were in play here, both of them encapsulated in the term "feudalism." One was a matter of the balance of power within states. Rulers confronted the entrenched might of the landed military aristocracy under conditions in which urban society was initially poorly developed; indeed, until the rise of towns the only substantial fiscal resources to which rulers had effective access were likely to be their own estates. The other theme was a characteristic idiom of political life, one that set tremendous store by rights and privileges of diverse origins, both individual and corporate. This idiom gave political conflict in the region a rather distinctive texture. A striking example of this, at a national rather than a local level, is the institution of parliaments. Rulers everywhere take advice, and it tends to be more or less understood from whom they will take it, and for whom, if anyone, the advisers can speak. But in later medieval England, for example, Parliament was a lot more than this. It was an institution through which rural and urban collectivities were formally represented; the ruler could not impose a new tax or make a new law without its consent, and his officials could be impeached by it. The ensuing conflicts between rulers seeking to do what rulers do everywhere and subjects bristling with rights and privileges could have a variety of outcomes. The ruler might be hamstrung, as was to happen in Poland; or he might find ways to bypass the thicket of rights and establish an absolute monarchy, as was to happen in France. But the most significant of the possible outcomes was something more like a tense symbiosis, as in the case of England.

The centuries following the turn of the millennium were also a period in which northwestern European culture, especially that of the nuclear regions of the old Frankish state, became more of a force to be reckoned with than in the past. There are two major contexts to look at. One is the traditional world of learning. What happened here did not burst the robust ecclesiastical framework that had preserved the culture of the late Roman world for so many centuries, but it greatly expanded the repertoire of this tradition. Learning acquired a new institutional focus, the university, and a new intellectual style, scholasticism, marked by a mercilessly systematic approach to exposition and argument. Meanwhile, the limited corpus of inherited Latin literature was enriched by a transfusion of previously unavailable material from Arabic and Greek.

By the fourteenth century a Muslim scholar could report a rumor to the effect that the philosophical sciences were widely cultivated in the land of the Franks, with abundant teachers and numerous students.

The other major context of cultural change in this period was the lay aristocracy. North of the Alps the collapse of the Roman world seems to have engendered a society in which the normal condition of the aristocracy was illiteracy—a stunning lapse by the standards of the Eurasian civilizations of the day. Even Charlemagne, an outstanding patron of literary culture, never really learned to write; his biographer tells us that this was not for want of trying, but that the king had come to it too late in life to make much progress. From around the thirteenth century this situation was changing. It was not that the lay aristocracy became proficient in classical Latin, as the Chinese gentry were in classical Chinese. Instead, what emerged was a compromise: French, the vernacular that had become the international language of the European aristocracy, became a literary language, while aristocrats (and not only aristocrats) became literate in it. Poetry in French already had a considerable history, but the development of a substantial prose literature was new. Much of the content of this literature was chivalric. But as early as the thirteenth century it was possible for readers literate in French to consult an encyclopedia; by the end of the fourteenth they could read works of Aristotle in French translation, and substantial libraries of books in French were to be found at many royal courts. Partly by a process of imitation and emulation, comparable vernacular literatures developed elsewhere in western Europe. The appearance of printing in Germany in the mid-fifteenth century, and its rapid spread, gave further impetus to these developments.

The Role of Southern Europe

Such indications of the increasing salience of the northwest did not mean that the south no longer mattered. Italy, or most of it, had remained Christian territory throughout the Middle Ages. It was the seat of the papacy, and thus the ecclesiastical metropolis of western Christendom. It was also blessed with a richer survival of Roman culture than the north; here, if anywhere, there may have been an unbroken continuity of learning outside the ecclesiastical context. Moreover, an important aspect of the political fragmentation of Italy was the reappearance of city-states, some of them aggressive mercantile and naval polities more reminiscent of Carthage than of Rome. All this led to the Italian Renaissance of the fifteenth century, with its extravagant revival of the cultural heritage of antiquity in a context far more secular than the universities of medieval Europe. Northwestern Europe proved very receptive to the Renaissance, but there is little reason to think that the region would have come up with it on its own.

Christian Spain, eclipsed by the long centuries of Muslim rule, had been of much less consequence than Italy for the cultural history of northwestern Europe, despite its leading role in the translation of Arabic texts. Yet it was from the Iberian Peninsula that in the late fifteenth century the expeditions were dispatched that rounded Africa and discovered the New World. Here too the northwestern Europeans took notice and soon became participants; but again, they showed no sign of being about to initiate the process for themselves.

We have nevertheless reached a point at which it no longer makes any sense to think of northwestern Europe as a provincial extension of the Mediterranean world. As we have seen in several connections, it was now a region in which major historical developments might have their beginnings. There was a further example of this in the early sixteenth century, the Protestant Reformation. This movement began in Germany and established itself widely in northwestern Europe. It thereby brought about a dramatic change in the landscape of western Christendom, irreversibly shattering its religious unity and unleashing a prolonged outburst of war and fanaticism. It also played a significant role in weakening the institutional and intellectual grip of the highly organized church that had been so central to the cultural life of the Middle Ages. Reforming the church meant, among other things, stripping it of most of its economic assets; as a Chinese emperor had already discovered in the ninth century, dissolving monasteries was a good way for a ruler to get rich.

II. The Monstrous Regiment of Women

John Knox (d. 1572) was the leading figure in the Protestant Reformation in Scotland, and did a good line in fanaticism. He was also the author of a tract of 1558 to which he gave the famous title *The First Blast of the Trumpet against the Monstrous Regiment of Women* ("regiment" in the sense of "rule"). Indeed, his two roles were closely connected. A key factor in determining whether the Reformation was adopted or rejected in any given region of western Christendom was the attitude of the reigning monarch, and in Britain at this time the rulers of both England and Scotland were faithful to the Catholic cause. By a strange coincidence they also happened to be women: Mary Tudor ruled England from 1553 to 1558, at a time when Mary of Guise was regent of Scotland on behalf of her ill-fated daughter Mary Queen of Scots—"our mischievous Maries," as Knox called them. So Knox, in exile on the Continent, sought to further the Protestant cause with a thunderous denunciation of female rule. His timing, however, was unfortunate, and the result was perhaps the greatest public relations disaster of the sixteenth century. Within months Mary Tudor had died, and God, moving in his mysterious ways, had saved England from popery through the accession of her Protestant sister Elizabeth. Knox's views on the monstrosity of the regiment of women had thus become an embarrassment to the cause, and his fellow reformers found themselves controlling the damage as best they could.

Kingdoms were supposed to be ruled by kings, and usually they were. Yet in many societies the vagaries of dynastic reproduction and court politics would issue from time to time in the rule of a woman. Even China, a land of impeccable patriarchal credentials, was ruled from A.D. 690 to 705 by the empress Wu. There was likewise a Hindu queen who ruled in southern India in the thirteenth century A.D., a Muslim one in Yemen in the twelfth, and so forth. The long history of Egypt boasts several female rulers, such as Hatshepsut in the fifteenth century B.C., Cleopatra in the first century B.C., and Shajar al-Durr in the thirteenth century A.D. Medieval Europe, too, had queens who ruled in their own right, and probably more than its share.

Yet, in one way or another, it was usually clear that such situations were felt to be anomalous. Reigning queens and their supporters were often visibly on the defensive. In some cases we see a tendency for female rulers to be referred to in the masculine gender, or be given masculine titles, or wear items of male clothing; in others their rule was justified by arguments to the effect that they were only superficially female. Thus a spurious interpolation in a Chinese translation of one of the Buddhist scriptures presents the womanhood of the empress Wu as just an inessential accompaniment of her (or his) current rebirth: "in reality you will be a Bodhisattva," the Buddha predicts to him (or her) in ancient times, but "you will manifest a female body." By the same token, the enemies of such queens had an easy time of it. When Shajar al-Durr became queen in Egypt, the caliph in Baghdad sent this message to the Egyptians: "If you haven't got a man left to appoint as your ruler, just tell us, and we'll send you one." Thus female rule might be an occasional fact of life, but to a greater or lesser extent it went against the grain of every major tradition of Eurasian civilization. "The country where a woman, a child, or a gambler rules sinks helplessly as a stone raft in a river," or so we read in an Indian epic.

So when Knox set about marshaling the misogynist resources of European civilization, they were there for him in abundance.

On the one hand, he had the strong support of the Christian Bible. Did not God impose a special punishment on woman for her egregious role in the fall of man, telling her that "thy will shall be subject to thy man, and he shall bear dominion over thee" (Genesis 3:16)? And did not the Holy Ghost, speaking through the mouth of Saint Paul, say, "I suffer not a woman to teach, neither yet to usurp authority above man" (1 Timothy 2:12)? And on the other hand, he could invoke the weighty authority of Greece and Rome. Aristotle had gravely underlined the pernicious effects of the undue influence exerted by the women of Sparta over their menfolk; was there any real difference, he had asked, between the rule of henpecked men and the rule of the women themselves? What, then, would Aristotle have said of "that realm or nation where a woman sitteth crowned in parliament amongst men"? Knox likewise referred his readers to the great codification of Roman law prepared under the emperor Justinian in the sixth century A.D.; there we read that "women are removed from all civil and public office." Since the Christian church represented the confluence of these Mediterranean traditions, Knox had no trouble finding "godly writers" whose fulminations he could adduce to support his case. Saint Augustine (d. 430) provided him with a rhetorical question that sums up the "uniform consent" of these learned authorities: "How can woman be the image of God, seeing she is subject to man and hath none authority, neither to teach, neither to be witness, neither to judge, much less to rule or bear empire?" And on top of this roll call of dead Mediterranean males, Knox assured his readers that female rule was "repugnant to nature."

There were nevertheless a few problems here and there. One that Knox faced explicitly concerned the Biblical heritage: had not Deborah been both a prophetess and a judge in ancient Israel (Judges 4:4–5)? Knox had already prepared for this awkward moment. After remarking of "all women" that "their sight in civil regiment is but blindness, their counsel foolishness, and judgment frenzy," he had paused to make an exception for females "such as

God, by singular privilege and for certain causes known only to himself, hath exempted from the common rank of women." Deborah was a case in point; with her, he had to admit, God had worked "potently and miraculously." But it would be quite wrong, he argued, to extrapolate from such exceptional cases "a tyrannical and most wicked law" permitting women to rule.

A problem Knox failed to confront in his tract arose from the Greek heritage. Aristotle's teacher Plato (d. 347 B.C.) was the author of the *Republic*, a discussion of the ideal state that was widely available in print by the middle of the sixteenth century; in this work he had placed in the mouth of his own teacher, Socrates (d. 399 B.C.), a view of women unusual in Athens at the time. Obviously, Socrates concedes, men tend to be better at most things than women; but this does not prevent some women being superior to some men. So there is no reason why, in the ideal city, women should be categorically excluded from any social role. Of course, it would take time for people to get used to the new order; much mockery would initially be directed at women being educated, bearing arms, and riding horses—not to mention exercising naked along with men in the wrestling schools. But then, as Socrates points out, it was not so long since the Greeks had thought it shameful even for men to be seen naked; yet in due course laughter died down and reason prevailed. So it would be with women. A woman, he argues, may have a natural talent for medicine, music, warfare, athletics, or scholarship; and by the same token, he concludes, she may be fitted by nature to become one of the guardians of the city. Thus the Greeks, like God, did not speak with a single voice.

Knox also omitted to discuss the testimony of ethnography, though this might well be thought relevant to his claim that female rule was repugnant to nature. Thus Tacitus had commented favorably on the unusual pattern of relations between men and women among an ancient people of northern Europe, the Germans. Though the women are somewhat skimpily clad, he tells us, both sexes observe a strict morality. They do not hasten into marriage, and when they do marry, the woman becomes the partner of the husband in all his toils and perils. In fact the men, he says, do not scorn to ask women for advice and do not lightly disregard it when given. Tacitus' account of the Germans was rediscovered in 1455, and created a considerable stir—not least in Germany, where it played to nationalist sentiments. By 1558 it had been printed many times. Indeed, Tacitus was called to witness by an Italian feminist who published a tract on the superiority of women over men in 1600 (naturally she also invoked Plato). But Knox did not mention Tacitus. Still less did he engage the ethnographic information on the peoples of the extra-European world that was beginning to accumulate in the Europe of his day.

In addition, Knox dismissed rather hastily the record of female rule in European history. A refutation of his tract that appeared in 1559 made much of this, pointing out with numerous examples that "women have reigned and those not a few" and further that "many countries have been well governed by women." Medieval European society routinely expected noble and royal women to stand in for absent husbands, or exercise power on behalf of minors, sometimes for long periods; Penelope at her loom would not have been much of a role model for them. Moreover, women occasionally inherited thrones in their own right (one reason for this being the rather unusual monogamy of European royal families, a practice that increased the risk of a king leaving only a daughter to be his heir). This bothered people more, but where misogyny and the principles of hereditary succession collided, misogyny tended to give way. It was a situation that all concerned had to make the best of; as one twelfth-century churchman told Queen Melisende of Jerusalem, "Although a woman, you must act as a man."

The aspects of the European heritage that Knox had to isolate, ignore, or brush aside should not blind us to the overall harmony between his views and the broader traditions of European civilization. But there was one very significant silence in his tract: nowhere in it did he employ the argument from segregation. In societies in which women—particularly elite women—are supposed to have no contact with men outside their immediate families, and to be kept away from the public sphere, it is hard to refute the view that the canons of decency would be violated if a woman, however talented, were to exercise political power. It was Knox's bad luck that there were no established norms of this kind in the society of northwestern Europe for him to invoke. This was part of a pattern of relations between men and women that Muslim observers had long found decidedly strange. In describing the Franks of twelfth-century Syria, one Muslim author commented on the freedom with which a Frankish wife out walking with her husband would chat to another man while the husband waited patiently for the conversation to be concluded. When an Ottoman ambassador sent to France in 1720–21 compared Paris to Istanbul, he remarked that Paris seemed much more populous than it really was; he explained that the women were always wandering about in the streets instead of staying at home, and that it was this mixing of men and women that made the center of the city appear so crowded.

In fact, northwestern European attitudes to the mixing of the sexes were not particularly unusual by ethnographic standards—they could readily be matched from West Africa, for example. It was in the context of the great civilizations of Eurasia that they stood out as exotic. What perhaps distinguished northwestern Europeans was the extent to which they retained their uncivilized attitudes even after they had otherwise become civilized. Giving women access to public space does not in itself give them power, but it undoubtedly eases the way for it, and makes the task of polemicists like Knox a little harder.

All this ultimately goes back to the obvious point that humans share the mammalian pattern of sexual reproduction. In surveying their history, we have taken this for granted. In all societies with a future, humans are to be found in two biologically distinct forms; and with few exceptions they belong unambiguously to one form or the other. Only women get pregnant, and only men are able to get them so. The straightforward biological differences between the sexes are at the same time clearly related to more diffuse behavioral traits; for example, no one seriously doubts that in general men have a stronger disposition to violence and are more given to taking risks in competing with each other. But if men and women are different, they are not all that different. Outside their immediate roles in sexual reproduction, there are few things that one sex can do and the other cannot. In contrast to the dithering that goes on in modern armies, full-blooded female warriors are well attested in the past as far afield as West Africa, the Near East, and China. All cultures thus have to live with both the reality and the limits of the difference. In making their rules, they might have had an easier time of it had men and women been either much more alike or much less so than they actually are. As it is, cultures can play up the difference, as in most of the civilizations of Eurasia, or they can play it down, as has increasingly been the tendency in the modern West. Either way, they usually end up, like Knox, in some kind of muddle. But underlying the variety, the confusion, and the indignity, striking regularities have obtained across cultures: at least in the public sphere, men have always tended to hold a disproportionate share of the power, and their competition has played a commensurate role in driving the course of history. This is why in earlier chapters of this book we did not stop to be puzzled by the fact that rulers are usually male, that the one true god of the monotheists was traditionally referred to by all and sundry as "He" and not "She," that only male Brahmins transmit the Vedas, that men play the dominant role in the Chinese ancestor cult, that respectable women do not attend elite Athenian drinking parties, and so forth. It is just more of the same.

Consider, for example, a move that the Aranda—that is to say, Aranda men—would sometimes make in intergroup diplomacy. If one group wanted to reach an accommodation with another, it would leave some women outside the other group's camp; if the men of the camp proceeded to enjoy the women, they thereby committed themselves to what was asked of them. This practice may surprise and even shock us, but it is a distant cousin of the more decorous use that kings used to make of their daughters, contracting royal marriages for them in order to cement alliances with other rulers (American presidents, by contrast, are not expected to treat their daughters as a diplomatic resource). But we would have been far more surprised (perhaps some of us pleasurably) had it been the women of the Aranda choosing to make their menfolk available to further the diplomatic interests of the group. And yet what men can do, women can do too: today at least we have some female ambassadors. We also have a good many female scientists, something conspicuously lacking in one of the more momentous European developments of the sixteenth and seventeenth centuries, the scientific revolution.

III. Jupiter's Paramours

The large-scale structure of the universe is in some ways easily accessible to human observation, even without the aid of advanced technology. The universe is transparent and contains numerous sources of visible light. As a consequence we receive a lot of information; it is not as if we were in the middle of a dense cloud or a dark cave. This means that it is possible to track the apparent motions of a variety of celestial bodies with great precision, and the results can be used to build a mathematical astronomy that predicts their future positions with considerable accuracy. Since the heavens are wide open to anyone with good eyesight (bar the modern inhabitants of light-polluted cities), some level of folk astronomy has been widespread, if not universal, in human cultures. Civilizations, with their greater potential for dividing labor and maintaining records, have tended to do considerably better.

Thus each of the Old World civilizations possessed a reasonably advanced astronomy, and they showed little hesitation in borrowing extensively from each other; in the late Middle Ages, for example, both Europe and China were heavily indebted to the astronomers of the Islamic world.

This traditional Old World astronomy may strike us today as primitive, but in early modern times it had by no means reached the end of the road. The most interesting question that could be put to it concerned the physical model that best accounted for the observed motions of the heavenly bodies. One issue here was which body lay at the center of the universe (assuming that it had one). In ancient Greece there had been two rival theories. According to the mainstream view, held by such authorities as Aristotle in the fourth century B.C. and Ptolemy in the second century A.D., the earth was at the center. The minority view, advanced as a hypothesis by Aristarchus of Samos in the third century B.C., placed the sun there. This ancient dispute was revived in 1543 when the dying Polish astronomer Copernicus finally published a book in which he championed the heliocentric view. He thereby gave this ancient hypothesis a renewed currency, but cannot be said to have proved it correct. It was the German astronomer Kepler (d. 1630) in the early seventeenth century who used the latest naked-eye observations to show that the heliocentric view was not just an alternative to that of Ptolemy but a vast improvement on it. His key insight was that the orbit of a planet is an ellipse with the sun at one focus. For this there was no ancient precedent; traditional astronomy, whether geocentric or heliocentric, had thought in terms of circular orbits, which resulted in much complication. But Kepler's laws of planetary motion were not just more elegant than the traditional system; they also predicted better, and from this point on the victory of heliocentrism was just a matter of time. Copernicus and Kepler had shown that naked-eye astronomy could still achieve a dramatic breakthrough.

Could astronomy have developed much further without a fairly dramatic change in the level of information available to it? In one respect it could have, and arguably did. It was possible to develop an understanding of the laws of motion on earth and then to apply them to the motions of the heavenly bodies—to explain not just *how* they moved, as Kepler had done, but *why* they moved as they did. This was one aspect of the work of the Italian scientist Galileo (d. 1642), and in 1687 it found its consummation in Newton's theory of gravity. But the other major aspect of Galileo's work was his brilliant exploitation of a recent technological innovation. The ability to make and shape glass had already been developed in ancient times, and with it some awareness of its optical properties. Starting from the thirteenth century there was talk of the use of lenses to make distant objects appear closer. But for practical purposes the telescope seems to have been a Dutch invention of the early seventeenth century. Galileo heard about it and began making telescopes for himself, so in late 1609 and early 1610 he was able to turn his rough-and-ready instruments on the heavens. The result was a cascade of simple, but momentous, discoveries that had hitherto been beyond the reach of even the most sophisticated astronomers. He published them in the spring of 1610, and they caused an instant international sensation—another testimony to the effect on European culture of the development of printing.

One of Galileo's discoveries concerned the well-known planet Jupiter. When he turned his latest telescope on it in January 1610, the planet itself had no secrets to reveal to him, apart from the fact that—like the other planets—it appeared in his telescope as a globe (see figure 22). He did notice three stars in the immediate neighborhood of Jupiter, two to the east and one to the west (January 7; east is to the left). But this did not particularly surprise him, since one of his major discoveries was that there were vast numbers of fixed stars that previous astronomers had been unaware of. His curiosity was nevertheless piqued by the fact that Jupiter and the three stars close to it were strung out almost exactly on a

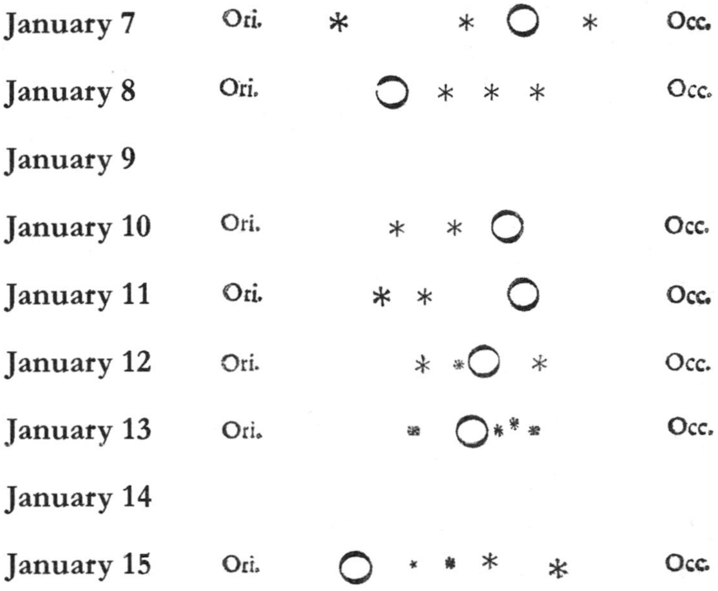

January 7	Ori.	* * O *	Occ.
January 8	Ori.	O * * *	Occ.
January 9			
January 10	Ori.	* * O	Occ.
January 11	Ori.	* * O	Occ.
January 12	Ori.	* *O *	Occ.
January 13	Ori.	* O* * *	Occ.
January 14			
January 15	Ori.	O * * * *	Occ.

Fig. 22: The Galilean satellites of Jupiter as observed by Galileo.

straight line. The next time he looked, he again saw three stars close to Jupiter, but this time they were all to the west (January 8). The obvious explanation was that Jupiter, against the background of the fixed stars, had moved a little to the east—except that according to the computations of the astronomers, the planet should at this moment have been moving west. Next time he saw only two stars, both now to the east of Jupiter, but as always in the same straight line (January 10).

At this point it was clear to Galileo that the movement of Jupiter could not account for the changing constellations he was seeing; amazingly, it had to be the stars themselves that were moving. After one further observation, Galileo concluded that far from being fixed stars, these objects were in fact revolving around Jupiter in the same kind of way that the planets Mercury and

Venus revolve around the sun. Later he confirmed that as Jupiter moved against the background of the fixed stars, it took its companions, or satellites, with it. In the meantime he had noticed something else: a fourth companion was sometimes visible (January 13 and 15). More or less correctly he took all four to be moving in circular orbits. He could infer from what he had seen that some orbits were larger than others and that the satellites closest to Jupiter had the shortest periods. But he was not yet able to determine the actual period of each satellite, since they looked so much alike; it was to be another two years before he had solved this problem. Nor did he name the individual satellites. It was a German rival who called them Io, Europa, Ganymede, and Callisto, after four of the paramours of the god Jupiter in Greco-Roman mythology (three of the four were females and one, Ganymede, was male). Giving Jupiter's celestial companions such names was a naughty, but erudite, Renaissance joke.

Galileo's discovery of Jupiter's four satellites was a stunning achievement, as he was well aware—he was not known for his modesty. It was also grist to his mill: he was, or soon became, an audacious and outspoken champion of Copernican heliocentrism against the traditional geocentric world system. One aspect of Copernicanism that had bothered people was a disturbing anomaly in its model of the solar system. The planets, in the Copernican view, all revolved around the sun; the moon alone persisted in revolving around the earth. This seemed at least untidy. Galileo's discovery now showed that, as a satellite revolving around a planet, the moon was in good company—a point he had already made when he reported his observations in 1610. In fact, his use of the telescope had provided him with a whole array of discoveries that could assist him in demolishing the geocentric worldview, and he used them with flamboyance and flair.

This eventually got him into trouble with the Catholic church. In 1616 Galileo, who now enjoyed celebrity status, was well received in Rome, but his heliocentrist arguments had the effect of drawing the attention of the censors to the book Copernicus had published in 1543; it was duly banned, and Galileo was ordered not to propagate its doctrine. This did him no personal damage, and in 1624 he was back in Rome being feted by the pope. But in 1632 he published his *Dialogue on the Two World Systems*, a superb vindication of heliocentrism. It was only then that things got nasty: he was put on trial for violating the terms of the injunction of 1616, and forced to recant his views under threat of torture. Had Galileo been a bit more restrained, or the church a bit more relaxed, this famous trial might never have taken place. After all, it had taken the censors almost three quarters of a century to notice Copernicus.

But if the trial was not inevitable, neither was it a fluke. In the course of the long centuries between the collapse of Roman civilization and the onset of the Renaissance, it was above all the church that had acted as the guardian of what survived of the literate culture of antiquity. In the process the church had invested heavily in the mainstream Aristotelian heritage that Galileo was now subverting with such abandon. But the affinity between the church and geocentrism was more than just a matter of long association. Geocentrism meshed with the unthinking assumption of most traditional cosmologies that the human race matters enough to have a place at the center of the universe. Even before the emergence of modern science, not everybody thought in this rather egocentric way. In the first century B.C. the Roman poet Lucretius, a zealous propagandist for the Greek materialist philosophy known as Epicureanism, had dared to assert that "not for us and not by gods was this world made; there's too much wrong with it." This assertion was then faithfully copied by the medieval monks to whom, paradoxically, we owe the survival of Lucretius' virulently antireligious polemic; but it was dead letter, despite the fact that the poem was printed again and again from the 1470s onward. Now, in the wake of the revival of heliocentrism, such views began to reappear, enhanced by the further realization that the sun was itself no more than a common star. "We cannot doubt," wrote the French philosopher Descartes (d. 1650), "that an infinitude of things exist, or did exist though they have now ceased to do so, which have never been beheld or comprehended by man, and have never been of any use to him."

To return to the corner of the cosmos that is our modest concern in this chapter, there is a geographical aspect of the story that is worth underlining. Galileo's life revolved around the cities of northern Italy, a region that had long been the most highly urbanized in Europe. But the future of astronomy, and of the scientific revolution in general, lay in northwestern Europe, above all in the Low Countries, France, and England. The condemnation of Galileo in Rome may have been one reason why astronomy now flourished in countries where the authority of Rome was either weak or nonexistent. Indeed, in Protestant Europe the persecution of Galileo by the Catholic church guaranteed him the status of a martyr and gave added prestige to his ideas. But the other reason why astronomy moved north, and perhaps the main one, was that northwestern Europe was finally coming into its own.

PART FOUR

TOWARD ONE WORLD?

CHAPTER 12

ISLAMIC CIVILIZATION

I. From Sea to Shining Sea

Our survey of the Eurasian landmass has left two major gaps (as well as numerous minor ones). The first is geographical. The regions we have covered represent the heartlands of the classic civilizations of Eurasia, but they do not include the extensive areas that remained outside them, or did so until recent centuries. There is a thick band of such unexplored territory to the north of the civilizations, and a certain amount of it to the south. The other gap is thematic. What, if any, were the prospects that the civilizations of Eurasia would be brought together into a more tightly integrated structure? Could Eurasia in any sense be united and, if so, how? The two gaps are very different in nature, but there is a surprisingly close linkage between them. The place to start is the question of Eurasian unification.

As we have seen, the various Eurasian civilizations had relations with each other even in ancient times. Shang chariots imply some kind of contact with the Near East; the alphabet spread outward from the Near East to the shores of the Atlantic and the Pacific; a

Roman medallion of the second century A.D. found its way to what is now Vietnam; paper made its way from China through the Islamic world to Europe. Examples of this kind can easily be multiplied, and they became more frequent as time passed. But this was cross-cultural seepage; it was not a flood capable of sweeping away the profound differences that separated the civilizations of Eurasia and gave them their distinct identities. Perhaps integration by cultural osmosis might eventually have come about had history had the patience to wait long enough. But of course it did not.

An altogether more drastic way of changing cultural landscapes is conquest. What, then, were the prospects for the cultural unification of Eurasia by the sword? The obvious thing to look for here would be a single Eurasian civilization with the will and the way to conquer the others. But the military history of the classic Eurasian civilizations did not point in this direction. There was no lack of states of imperial dimensions built up through conquest, but for the most part they retained a firmly regional character. The major exception was the career of Alexander the Great in the fourth century B.C., which led to the imposition of a Greek elite culture on much of the Near East. But his conquests did not go much farther. His invasion of India did not amount to more than an incursion, though Greeks later ruled some territory in the northwest; China was beyond his horizon; and more surprisingly, neither he nor his successors ever conquered the western Mediterranean. No other conqueror who set out from any of the older civilizations of Eurasia was in the same league.

At first sight it makes even less sense to expect such a process to be initiated from the peripheral regions of Eurasia, with their much less complex societies. We take it for granted today that, other things being equal, rich and advanced societies make war more effectively than poor and primitive ones. But Cyrus, the founder of the Persian Empire in the sixth century B.C., had a different view of the matter. "Soft countries," he said, "breed soft men; it is not the property of any one soil to produce fine fruits and good soldiers too." Cyrus drew his hard men from the rugged uplands of Persia, but that was not the only possibility.

The Northern Tribes

Harsh territory abounded across the top of Eurasia. In the far north the bleak, open wilderness of the tundra stretched in a thin strip from Scandinavia to Siberia; immediately to its south lay a wider band of forest, the taiga. This Arctic territory was populated mostly by hunters like the Yukagir of northern Siberia and by reindeer pastoralists like the Lapps of northern Scandinavia. Such peoples tended to let history pass them by. But between this northern world and the civilizations far to the south lay the Eurasian steppes, reaching Manchuria in the east and Hungary in the west. Like the uplands of Persia, this belt of grasslands did not bear fine fruits. But it was the homeland of the horse and doubtless the scene of its domestication. Already in the time of Cyrus the steppes were teeming with nomadic pastoralists. These pastoralists were hard men, and those whose economies were built around the horse had two further military advantages. First, they were nature's cavalry. A Latin author of the fifth century A.D. had this to say of one such people, the Huns, who at the time were invading the western Roman Empire: "Scarce has the infant learned to stand without his mother's aid when a horse takes him on his back. You would think the limbs of man and horse were born together, so firmly does the rider always stick to the horse, just as if he were fastened to his place: any other folk is carried on horseback, this folk lives there." Second, nomadic pastoralists were supremely mobile. Scholars dither as to whether or not to identify the Huns with the Hsiung-nu who threatened the northern frontier of China starting in the late third century B.C. But it is a matter of record that in the thirteenth century A.D. the Mongols attacked both Germany and Japan.

The earliest people of the steppes to have a serious impact on the outside world were probably the Indo-Europeans, though they were by no means exclusively pastoralists. In our survey of the regional civilizations of Eurasia we repeatedly encountered peoples speaking languages of the Indo-European family; already in ancient times such languages could be found as far afield as the British Isles and northeastern India, and the fact that they constitute a family means that they must go back to a single ancestral language—"proto-Indo-European." As we saw, some of the ancient expansion of the Indo-European languages took place as late as the first millennium B.C., but a good deal of it must be older. When and where did it start? One set of clues is linguistic. Though we have no direct knowledge of the ancestral language, it is possible to reconstruct some of its vocabulary by the systematic analysis of common elements in the vocabularies of the known daughter languages (screening out later loanwords in the process). Such reconstruction points to an economy in which farming was well developed, especially on the pastoral side, where it included the domesticated horse; wheeled vehicles were certainly in use, but such metallurgy as existed may have been limited to copper as opposed to bronze. When and where would this be? From an archaeological point of view the most diagnostic elements are the domesticated horses and the wheeled vehicles. The horse must surely have been domesticated somewhere in its natural range; this takes us to the steppes, where both the domesticated horse and the cart make their appearance in the archaeological record in the fourth millennium B.C., prior to the onset of the Bronze Age. Within the steppes there is reason to pick out the region to the north of the Caspian and Black Seas as the most likely Indo-European homeland. All this can only be a hypothesis, but it is undoubtedly the best available. What is satisfying about it is that it also provides us with a measure of explanation of the Indo-European expansion that must have followed in the third millennium. A people whose way of life includes horses and carts is mobile, and at the same time it has a military advantage over pedestrian peoples, whom it is liable to push aside or assimilate.

Yet, of all the steppe peoples of history, only the Mongols came close to creating a Eurasian empire. Their highly destructive conquests in the thirteenth century A.D. included large parts of East Asia, the Near East, and eastern Europe (the lack of steppe to the west of Hungary does something to explain their failure to push on into western Europe). In the end, however, their imperial venture came to nothing. The far-flung Mongol empire lasted only a few decades; already in the thirteenth century, it was divided into increasingly autonomous states, and most of these disappeared in the course of the fourteenth century. Just as significant is the fact that the Mongols had brought no civilization to spread among their battered subjects, tending rather toward eventual assimilation into the local milieu. In the end, the cultural map of Eurasia after the Mongols did not look very different from what it had been before.

There was, however, one thing the Mongols did for Eurasia. For a period of a few decades, after their initial conquests and before the disintegration of their states, the Mongols made it possible for people to travel from one end of Eurasia to the other. Thus the Christian monk Rabban Sauma, born into a Turkish people living on the northern frontier of China, found his way from Peking to Bordeaux, while the Venetian merchant Marco Polo traveled to China with his father and uncle and found employment in the service of the Mongol ruler there. This mobility was not an unmixed blessing—it probably helped to spread a vicious epidemic of plague from one end of Eurasia to the other in the middle of the fourteenth century. But it did provide an unprecedented opportunity for the civilizations of Eurasia to learn more about each other. In Iran, for example, it became possible in Mongol times for a reader of Persian to consult systematic accounts of the histories of China and western Europe (see figure 23; that everyone looks oriental reflects the strength of Chinese artistic influence in Iran in this period). We will come back to the implications of this widening of Eurasian horizons in the next chapter.

Fig. 23: ***Above:*** **Charlemagne (*right*) and Pope Stephen IV (*left*), from an early fourteenth-century Persian history of the Franks.** ***Below:*** **the empress Wu (*far right*) and her immediate successors, from a parallel history of China (the artist clearly did not know that the empress was a woman).**

The Southern Tribes

If the peoples to the north were not going to unify Eurasia, what about those to the south? Here the lay of the land looks even less propitious. South of the civilized lands of the eastern half of Eurasia lies the Indian Ocean, a significant commercial scene since ancient times, but not the stuff of which conquests are made. South of the civilized lands of the western half of Eurasia there is desert—notably the vast expanse of the Sahara. This was miserable territory for agriculturalists outside a few oases, but it did provide sufficient vegetation for a thin and hardy population of nomadic pastoralists. One Berber people of the western Sahara came out of the desert in the eleventh century to conquer much of North Africa and southern Spain; but they were Muslims like the peoples they conquered, and their impact was a transient one. That leaves us with Arabia, essentially a fragment of the Sahara cut off from the main body of the desert by the Red Sea. It hardly seems worth troubling with so limited and unpromising a territory in the present context. And yet the single force that did most to unify Eurasia before modern times came out of Arabia.

Before we proceed to the rise of Islam among the Arabs, we should go back to the spread of Christianity in the early Middle Ages. The crucial event, as we saw in chapter 10, was the adoption of the Christian faith as the state religion of the Roman Empire. This made it much more attractive, and not just inside the imperial boundaries. From the fourth century onward conversion became epidemic in the lands beyond the Roman frontiers, with peoples as diverse as the Irish, the Goths, and the Ethiopians jumping on the bandwagon. Of course there were holdouts. As late as 1128 the Pomeranians, a Slavic people of the Baltic coast, were still pagans. But significantly, a group among them was arguing "that it was incredibly stupid to separate themselves like miscarried children from the lap of Holy Mother Church, when all the provinces of the surrounding nations and the whole Roman world had submitted to the yoke of the Christian faith." Why, then, should the Arabs have seen things differently?

One thing that was different about Arabia was the poverty of its desert environment. Poverty militates against complexity. Pre-Islamic Arabian society was tribal, with a form of social organization based on extended kinship. It did not possess a wealthy aristocracy clearly differentiated from the masses, and still less did it have powerful rulers. This made the mechanics of conversion significantly harder. In the societies on the edges of the Mediterranean world, the king was normally the key actor in the process of conversion. Thus when Saint Augustine of Canterbury arrived in Kent in 597 to convert the English, he had neither the need nor the opportunity to involve himself in the micropolitics of the local clans. Instead, he found himself preaching to Ethelbert, the Kentish king. Ethelbert was no novice, having been on the throne since 560. He doubtless recognized a tricky combination of danger and opportunity, and wisely decided to play for time; but he established ground rules for Augustine's mission that opened the door to the rapid conversion of his people. It would have been hard to find a ruler with this kind of authority in Arabia.

Ethelbert had good reason to facilitate the adoption of Christianity in his kingdom. From the point of view of the pagan English, continental Christianity was civilization, and civilization has much to offer to a king—literate bureaucracy, for example. But adopting Christianity could not be cost-free, as Ethelbert knew very well. When he played for time in his initial audience with Augustine, he said this: "The words and promises you bring are fair indeed, but since they are new and uncertain I cannot assent to them, abandoning what I, with the whole English people, have observed for so long." There may be good reasons to break faith with the religion of one's ancestors and adopt a foreign one; indeed, it may even be incredibly stupid not to do so. But the process is necessarily jarring, and wounding to ethnic pride. This was Ethelbert's dilemma, and the Christian mission in Kent was by no means insensitive to it. The pope, after long deliberation, recognized that the stubborn English could not be expected to make an immediate break with everything in their pagan past, and endorsed a compromise whereby they could retain their temples intact once the pagan idols had been removed from them. So the English gave up their idols, but for a time at least they held on to their temples. This, incidentally, shows them to be reasonable people; some of the eighteenth-century Voguls—who lived in northern Russia and were distantly related to the Hungarians—were not willing to convert unless their ancestral idol was baptized with a gold cross and given a place in church beside the icons. But to return to the Arabs: the absence of a powerful king to take the lead in Arabia made it that much harder for them to accept the ethnic costs of conversion.

Would the Arabs, then, take the plunge and adopt Christianity, or would they remain obdurately pagan? Their unique historical importance arises from the fact that they did neither, instead adopting a monotheist religion of their own. The key figure in the formation of Islam was Muhammad. His career is narrated in great detail in the Islamic sources, though it is hard to say how reliably. Here is the gist of what they tell us. Muhammad's hometown was Mecca, a settlement in western Arabia inhabited by an Arab tribe; it was lacking in agricultural resources, but possessed a well-known sanctuary, the Kaᶜba. The tribe was divided into a number of clans. It had no chief, and it recognized no superior authority either in Mecca or elsewhere. Early in the seventh century A.D., Muhammad began to receive the revelations that were in due course to be embodied in the Koran, the Muslim scripture. His message was unambiguously monotheist, but it was neither Jewish nor Christian. Although he succeeded in making some converts in Mecca, his monotheist incivility with regard to the pagan gods caused great offense to the pagan majority. Muhammad personally was not in much danger, but his followers were in dire need of protection, and it was not obvious where this could come from. The king of Ethiopia was sympathetic, and Muhammad sent some of his followers to take refuge with him; but if Islam was to be a contender in Arabia, this king was too far away to help. Muhammad also sought to find a tribe closer to home that might be willing to extend its protection; he did this by attending fairs at which members of different tribes would gather to trade. But he had no success.

Finally he had his opportunity. The oasis of Yathrib (now Medina) lay some two hundred miles north of Mecca. It was much larger and had a much more complex population: two Arab tribes and three Jewish ones. Like Mecca, it was not under the sway of any chief or ruler; unlike Mecca, it was in political disarray. While attending a fair, Muhammad met some members of one of the Arab tribes and told them about Islam. They responded positively, partly because living alongside Jews had prepared them to recognize a prophet when they saw one, and partly because they perceived Muhammad as the solution to their political problem. When they had left Yathrib, they told him, its people were more riven by internecine feuds than any other; "maybe," they said, "God will unite them through you." They undertook to go back to their people and summon them to Islam, adding, "If God unites them in it, there will be no man more powerful than you!" And so it turned out. Or more precisely, after Muhammad and his followers moved to Yathrib in 622, he united the Arab tribes and eliminated the Jewish ones. By the time he died, in 632, war and diplomacy had made him the ruler of an Islamic state. All this was quite unlike the process by which other peoples of the day were converting to Christianity. Whereas Augustine had arrived in Kent to find a state already firmly in control, Muhammad had created one. And it was a state of a very different kind: he was not a king but a prophet.

How, then, did Muhammad negotiate Ethelbert's dilemma? His words and promises were fair indeed, but how could the pagans of Arabia adopt them without abandoning the age-old beliefs of the whole Arab people? Here we have to go back to the oldest monotheist scripture, the Bible. This substantial book speaks at length of many things, but it never so much as mentions such worthy peoples as the English, the Irish, or the Pomeranians. Yet it has something quite significant to say about the Arabs, especially if one knows (as everybody did at the time of the rise of Islam) that when God speaks of the Ishmaelites, it is the Arabs to whom he is referring. They were called Ishmaelites because according to Biblical genealogy they descended from Ishmael, a son of the patriarch Abraham—unlike the Israelites, who descended from a different son of Abraham, namely Isaac. The question, as often in such situations, was which son would get what, and the answer was provided by God in a conversation with Abraham. He promised to bless Ishmael, and to make of his descendants "a great nation"; but, he insisted, "my covenant I will establish with Isaac" (Genesis 17:20–21). The Arabs were therefore a sideline, albeit a relatively favored one. Thus far the Bible.

Many centuries later Islam told the same story, but told it differently. In the Koran we read of Abraham and Ishmael building God a house; the Bible makes no mention of this house, but Muslim tradition identifies it as the Kaᶜba in Mecca. As they work, they pray to God to make of their seed a nation submissive (*muslim*) to him, and eventually to send among them "a messenger, one of themselves," to instruct them and purify them (Koran 2:127–29). The Arabs, then, were destined to be more than just a great nation. The Kaᶜba, though it had come to look like a pagan sanctuary, had originated as a monotheist temple; no wonder Islam had no problem retaining the pre-Islamic pilgrimage to Mecca in an Islamized form. It was as if the English could have sanctified a pagan temple in Canterbury as an Abrahamic foundation. At the same time Muhammad was manifestly the prophet for whom Abraham had prayed, and an Arab like the people to whom he was sent. The English owed their evangelization to a monk from Rome, while the Irish Americans who march in New York on Saint Patrick's Day celebrate the conversion of their ancestors at the hands of a British slave. Muhammad, by contrast, was an Arab, and sent with an "Arabic Koran" (Koran 12:2). In short, adopting Islam did not mean abandoning the age-old beliefs of the whole Arab people; all that had to be discarded was the pagan overlay that had come to obscure the pristine monotheism of their ancestors. For an Arab, to convert to Islam was to be true to oneself.

The Islamic World

What Muhammad had done was remarkable, but it hardly affected anyone outside Arabia. That changed soon after his death. Under his successors, the caliphs, the Arabs began to attack the Fertile Crescent, the region adjoining Arabia on the north. They had always been in the habit of conducting small-scale raids there, but the new state enabled them to coordinate their efforts in an unprecedented way. In the course of the 630s and 640s, these attacks developed into a series of conquests that included the entire Persian Empire and most of the southern provinces of the surviving eastern half of the Roman Empire. The Muslims subsequently extended their rule all the way along the coast of North Africa, and in the early eighth century they went on to conquer most of Spain; meanwhile, in the east they advanced into Central Asia and northwestern India.

No previous empire in history had extended over such a distance, and well into the ninth century most of these territories were held together under the rule of a single state, the caliphate. The history of this state was turbulent, with numerous rebellions, several periods of intense civil war, and a major change of dynasty in the mid-eighth century; moreover, the political values that emerged in Islam were by no means as state friendly as those of China's Confucians (recollect the ninth-century Muslim who wept at the thought of his father seeing him serving the caliph as a judge). But the duration of this effectively united caliphate was far longer than that of the Mongol empire, and the Muslims had something to which the Mongols would offer no parallel: a novel monotheist religion. The outcome was that a new civilization took shape around Islam, and by the time the caliphate broke up, this civilization was firmly established over lands that had previously displayed quite different cultural allegiances. There was large-scale conversion of non-Arabs to Islam; except in the southern fringes of Europe, the non-Muslim population was gradually reduced to scattered minorities that no longer threatened Islamic dominance.

In the Fertile Crescent, Egypt, and, eventually, most of North Africa, the mass of the population also became Arabic-speaking, and their pre-Islamic languages fell into disuse. At the same time a new elite culture was established, centered on the Islamic religion and the Arabic language; Arabic became the classical language of a civilization in the manner of classical Chinese or Latin, and everything that an educated elite might want to read became available in Arabic. In one sense Islamic civilization was not new: most of the raw materials of which it was made derived from the cultures the Arabs had conquered—which is why it was possible to bring a new civilization into existence with unique rapidity, and to an extraordinary extent as a result of the career of a single man. But the reshaping of the diverse materials yielded a civilization quite distinct from any of its predecessors, and one that replaced them over large areas.

After the caliphate fell apart in the ninth century, the lands of Islam were never again ruled by a single state. The nearest approach to a restoration of the original unity of the Islamic world came with the Ottoman Empire. This state had its beginnings in Anatolia, a territory brought into the Muslim world by Turks from Central Asia who conquered it in the eleventh century. The Ottoman state emerged in the late thirteenth century and lasted until the early twentieth. At its height, in the sixteenth century, it included the Balkans, Anatolia, the Fertile Crescent, parts of Arabia, Egypt, and North Africa as far as the borders of Morocco; at the same time, the activities of the Ottoman state extended as far afield as Mombasa and the Volga, Andalusia and Aceh. But even in this period there were at least two other Muslim states that could be described as empires. One was the Ṣafawid state in Iran, where the sixteenth century saw the establishment of Shīʿism, a form of Islam that is heterodox in relation to the mainstream Sunnī confession. The other was the Moghul empire, which represented the zenith of a further extension of Muslim rule over India that had begun in the eleventh century.

The inroads of Islam in India were primarily the result of conquest. But in several other areas the spread of Islam had more to do with the activities of Muslim merchants and the cultural choices of indigenous peoples. As we saw in chapter 6, there were two such regions in Africa: the West African interior south of the Sahara, and the East African coast. There was a comparable spread of Islam in parts of northern Asia. As early as the tenth century the Bulgars who lived around the bend in the Volga (cousins of the ones who gave their name to Bulgaria) were making well-meaning attempts to adopt Islam; a Muslim ambassador dispatched to them by the caliph did his best to wean them off an un-Islamic taste for nude mixed bathing in the Volga. In later centuries Islam spread among the tribes of the steppes to the east of the Volga, and farther south it became established in Chinese Turkestan. Meanwhile, the late medieval period saw a substantial penetration of Islam into parts of mainland and island South East Asia. By the middle of the sixteenth century there was a significant Islamic presence on the island of Mindanao, in the south of what we now call the Philippines; no doubt Islam would eventually have spread over the entire Philippine archipelago, had not the Spanish conquered it in 1565 and named it after their king.

Islamic civilization had thus become the prevalent culture from Morocco to Mindanao, from the Atlantic to the Pacific (see map 10). As such it was well on the way to dominating the Old World as a whole. There were, of course, extensive regions in which Islam was not established. They included most of southern Africa, particularly in the west, where there had been no coastal spread of Islam to match the ribbon development in the east; all of western Europe, one of the few regions in which Islam lost ground in the Middle Ages; vast tracts of northern Eurasia, including the Slavic continuation of the eastern Roman tradition in Russia; China and most of its neighbors, despite the emergence of a Chinese Muslim minority; large parts of Southeast Asia, such as Buddhist Thailand and Hindu Bali; and India, in the sense that the bulk of the population remained Hindu. All this added up to a lot of territory, but on the map of the Old World as a whole it appeared as disjointed fragments with nothing in common beyond the fact that they were not Muslim. Against this background it would have been easy to imagine a history in which Islam continued to spread for a few centuries in the Old World before making its way to the New World and the Antipodes. In the event, the outcome was very different, as we will see in the next chapter. But if we are looking for a potentially global culture in the centuries prior to the European expansion, then it is Islamic civilization that best fits the bill.

Was it accidental that a civilization centered on Islam played this role in the history of Eurasia? Or did the religion itself possess features—over and above the traditional intransigence of monotheism—that helped the civilization to achieve its remarkable expansion and maintain its cultural unity over unprecedented distances?

One relevant feature was the salience of holy war (*jihād*) in the Islamic heritage. The Koran, though not by any means a pacifist manifesto, is somewhat ambiguous in its treatment of armed struggle against the infidel. Some passages seem to enjoin only defensive warfare: "fight in the way of God with those who fight you, but aggress not: God loves not the aggressors" (Koran 2:190). Other passages, however, suggest aggressive warfare: "slay the idolators wherever you find them, and take them, and confine them, and lie in wait for them at every place of ambush" (Koran 9:5). Likewise Muhammad is quoted as declaring, "I have been commanded to fight people till they testify that there is no god but God and Muhammad is the messenger of God, and perform the prayer and pay the alms-tax." Yet, at one point in the course of his struggle with the pagan Meccans, he made a truce with them, and he is also said to have told his followers, "Leave the Turks alone as long as they leave you alone." On the basis of such materials, the medieval Muslim scholars worked out an elaborate legal doctrine of holy war. They endorsed the fundamental idea of aggressive

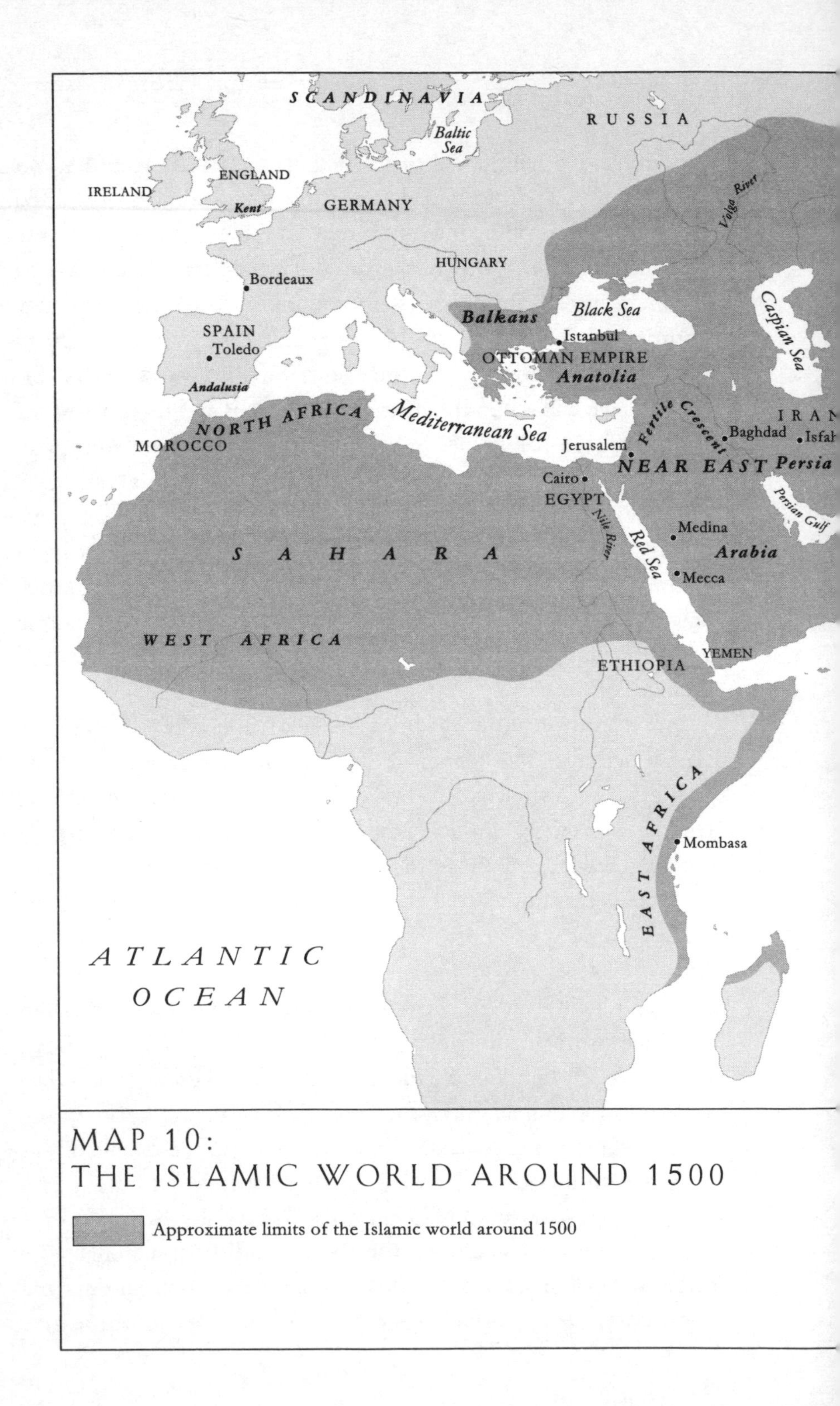

MAP 10:
THE ISLAMIC WORLD AROUND 1500

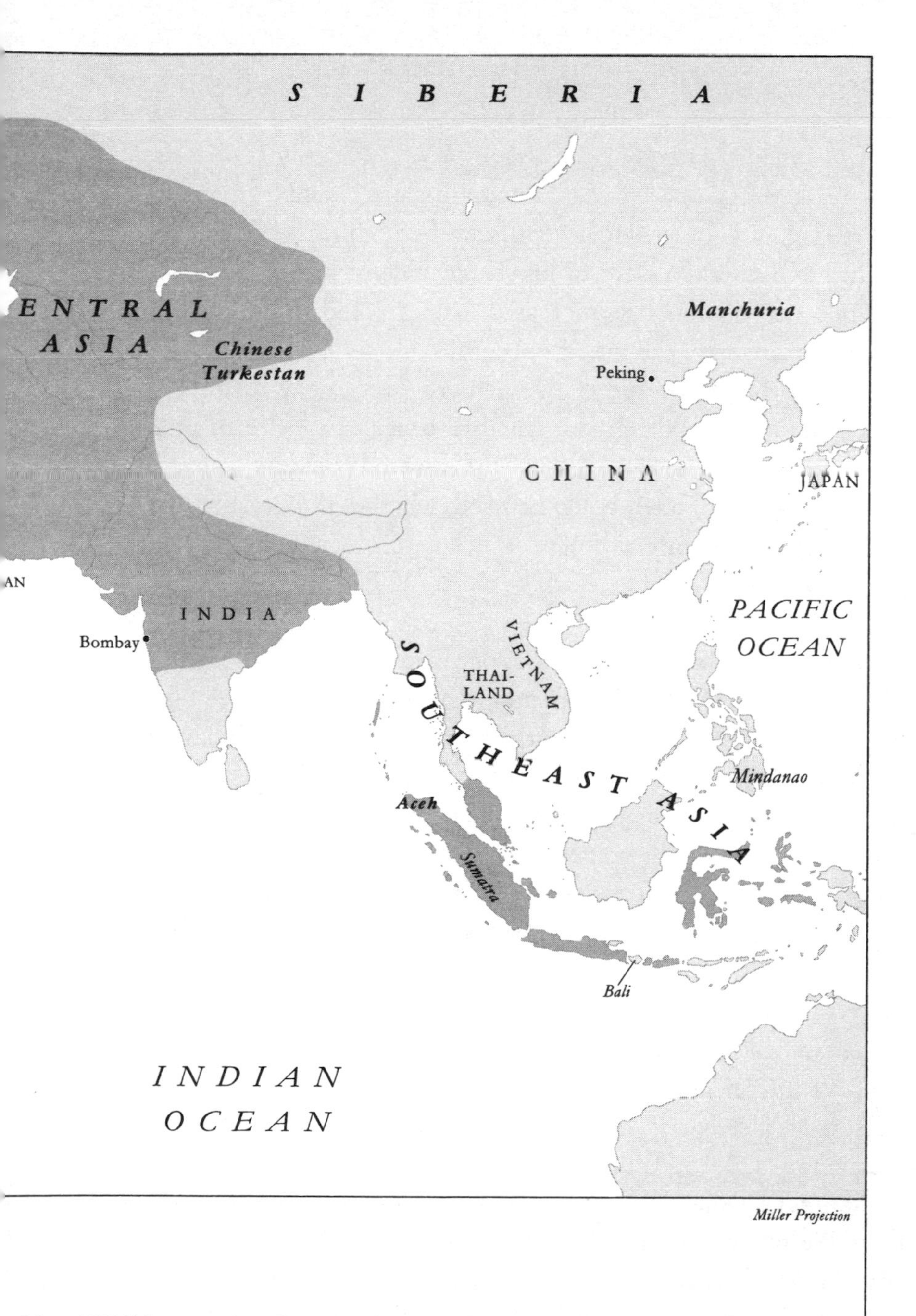

After 1500 Islam continued to spread, especially in sub-Saharan Africa and island Southeast Asia, where it reached Mindanao. However, the Islamic world has contracted in eastern Europe and India, in regions where despite prolonged Muslim rule the majority of the population remained non-Muslim.

warfare aimed at extending the dominion of Islam, but at the same time hedged it about with a variety of ifs and buts. Normally the infidels were to be called to accept Islam before they were attacked, women and children were to be spared, enemies were not to be tortured nor their bodies mutilated, truces could be made, Jews and Christians who submitted to Islamic rule were to be tolerated, and so forth. On the other hand, there were many matters on which the scholars disagreed: whether other categories of non-combatants could be killed, whether the enemy could be slain with fire, whether mangonels could be used, whether the livestock and fruit trees of the infidel could be destroyed, whether non-Arab pagans who submitted to Islamic rule could be tolerated, and the like.

The fact that Islam espoused a doctrine of holy war against the infidel was obviously no guarantee that Muslims would actually engage in it. Granted it had worked like a charm to mobilize the anarchic tribesmen of seventh-century Arabia. But humans regularly fail to live up—or down—to their principles. Nor is lacking such a doctrine any guarantee that people will abstain from conquest. The Mongols, for example, conquered more territory than the Muslims, and slaughtered many more people, without being inspired by anything that could pass for a doctrine. Yet the central place of the value of holy war in the Islamic heritage certainly made available to Muslims a moral charter for the continuing conquest of infidel lands, and one they invoked often enough. In that sense there clearly was something about Islam that lent itself to the creation of a global culture.

Another Islamic duty worth attention here is of a very different kind. Pilgrimages are commonplace in the world's religions. The adherents of a given faith will usually regard some places as more sacred than others. The reasons for this may vary; thus the place may have some cosmic significance, or have been the scene of an important event in the life of the founder, or be the tomb of a revered saint. What believers do about such places may also vary:

they may travel to them to join in crowded communal rituals, or visit them as individuals, or even make a point of avoiding them. This kind of thing is as widespread in Islam as in other world religions. What, then, is different about Islamic pilgrimage?

Of the many forms of pilgrimage in the Islamic world, one stands out: the pilgrimage to Mecca (*ḥajj*). There are several reasons why Mecca is a more significant place in Islamic terms than the average Arabian oasis, but the most fundamental is the one we have already encountered, the presence of God's house. In some religions gods have multiple residences, but in the Islamic case God's house is unique. Here, then, is a sacred place that has priority for all Muslims, just as Jerusalem does for all Jews. By contrast, it is not at all clear what single place would enjoy such a status for adherents of either of the other world religions, be they Christians or Buddhists. Moreover, going on pilgrimage to Mecca is not just an act of piety. It is an obligation of the individual believer laid down in the Koran: "It is the duty of people towards God to make the pilgrimage to the House, if he [the believer] is able to make his way there" (Koran 3:97). Again, the scholars went to work to produce a full legal account of the duty. A Muslim is obliged to perform the pilgrimage once in a lifetime (though there is nothing wrong with doing it repeatedly). Certain categories of people are exempt, such as slaves, lunatics, or unaccompanied women. Insecurity on the roads and lack of means excuse those who would otherwise be obligated (the Koran says "if he is able"). The scholars also spelled out the traditional rituals that the pilgrims must perform when they gather in the neighborhood of Mecca in the second week of the last month of the Muslim year. The net effect of all this was to confer on the pilgrimage to Mecca the status of a fundamental institution of Islam.

This meant that in Islam, to a far greater extent than in Christianity or Buddhism, there was a place and a time each year at which believers from the most diverse components of the Islamic community were gathered together. The Muslim world, like any other of its day, was one in which most people lived their lives within very narrow horizons. News traveled slowly across landscapes that were politically, ethnically, and culturally fragmented. In such a world the pilgrimage played an unusual role in creating a broader consciousness of the geographical range of Islam, and in making possible the maintenance of regular contact between widely separated Muslim populations. Mecca was admittedly far from perfect in this respect. At the best of times it was a difficult place to get to—pilgrims came at the risk of being robbed by the nomads in the desert or shipwrecked in the Red Sea. And when the pilgrimage fell in high summer, Mecca was a grim place for the pilgrims to be wandering around in the open. But as a way of keeping a civilization in touch with itself, it was peerless in premodern times.

No world religion can be monolithic: it is in the nature of such a religion to comprehend disparate people living diverse lives. But this does not mean that it has to be a heap of rubble. As might be expected, religions vary. For example, both Christianity and Buddhism are much more fragmented by sectarian divisions than Islam, both had significantly lower levels of internal contact in premodern times, and neither had the same degree of cultic uniformity as Islam. So if we want to place the world religions along the spectrum from heap of rubble to monolith, there is little doubt that Islam falls closest to the monolithic pole. There were nevertheless small, but quarrelsome, groups of Muslims for whom the Islam of premodern times was not monolithic enough. To them much of its local diversity was beyond the pale of toleration. Historically the most significant movement of this kind was Wahhābism, a fundamentalism that appeared in the Arabian interior in the eighteenth century in alliance with the Saudi state. At that stage the Saudis were able to impose Wahhābism over much of Arabia by waging holy war on adversaries whom they deemed to be infidels; but they lacked the military capacity to spread it farther. Later, however, this reformist movement was to strike a chord over much of the Muslim world.

II. Muslim Ethnography

It was the Greeks who created the first literature that seriously attempted to describe the manners and customs of other peoples. Of course, any body of literary texts is likely to contain incidental references to such matters, but that does not amount to ethnography. Nothing that could be called by that name survives from ancient Egypt or the ancient Near East. Nor did the other classical civilizations of the Old World contribute much to the field: the interest of the Chinese in other peoples was limited, and that of the Indians was minimal. Tacitus wrote his booklet on the Germans in Latin, but in such matters the Romans were the pupils of the Greeks. The ethnographic literature of the Greeks and Romans thus provides us with an unusually interesting record, and one that extends as far east as India. But it cannot compare in geographical range to what the Muslims could and did undertake. Today the primary language of ethnography the world over is English; in the Old World of medieval times, it was Arabic.

The Muslims inherited from their Greek predecessors a general theory of ethnology. It is clearly presented in a work by Ṣāʿid al-Andalusī, a Muslim scholar who served as judge in the Spanish city of Toledo in the eleventh century, shortly before it was reconquered by the Christians (he died in 1070). As a judge he must have been learned in Islamic law, but his consuming passion was science. He saw science as a cosmopolitan enterprise and wrote a short work setting out the contributions made to it by the various peoples of his world. His starting point was the fact—it was still more or less a fact in his day—that scientific achievement was a characteristic of the peoples of the temperate latitudes: the Indians, the Persians, the Chaldeans, the Greeks, the Romans, the

Egyptians, the Arabs, and the Hebrews. The other peoples of the world had not contributed to the development of science, though the Chinese and the Turks enjoyed a rank above the rest (the Chinese in particular were superior to all other peoples in craftsmanship, but mere technology did not count as science). The reason for this geographically restricted distribution of scientific talent was climatic. To the north of the temperate band the rays of the sun were too feeble, with the result that peoples living in high latitudes were blond and stupid. To the south of the temperate band the sun was too strong, so that peoples living in low latitudes were black and foolish. Those in between were just right, hence their glorious role in the history of science.

Two things stand out about this dumb-blond theory of ethnology. One is that its racism was environmental, not biological. Our author does not himself make this explicit, but it is clear from another presentation of the same theory—this time by the famous theorist of history Ibn Khaldūn (d. 1406)—that blacks who settle in the lands of the blonds will eventually turn blond themselves, and presumably vice versa; your destiny is in your latitude, not in your genes. The other thing to note is that the whole conception has nothing to do with Islam. We could well imagine a view of the world that regarded Muslims as civilized people, and everyone else as barbarians. But that is not what we see here. The list of scientific peoples consists overwhelmingly of non-Muslims; the Indians, for example, were known to be rank idolators. Yet our author regards these eight peoples as "God's elite." Conversely, by the eleventh century there were Muslims as far north as the bend in the Volga and as far south as Ghana; but he does nothing to protect them from the sweep of his generalizations. Only in his own corner of the world does he find himself forced to make exceptions, conceding that the Berbers and Galicians are utterly barbarous peoples. What is interesting here is the irrelevance to our author of the fact that the Berbers were Muslims and the Galicians Christians. His problem is simply to explain how such awful people could be found at so excellent a latitude; his answer is to shrug his shoulders and explain that God does what he pleases.

This solar theory was widely known in the Muslim world, but when Muslim authors provided concrete accounts of exotic peoples, they tended to forget about it. Thus the view that blacks were foolish played no part in the accounts of Ghana or East Africa referred to in chapter 6; indeed, the account of East Africa goes so far as to ascribe eloquence to native preachers among the Zanj—eloquence being a quality the Arabs normally valued too highly to associate with anyone other than themselves. But there is one venture in ethnography on which the theory sheds a certain light.

Bīrūnī was an older contemporary of the judge of Toledo (he must have died a bit after 1050), but he lived at the other end of the Muslim world. Vastly erudite in a range of fields, he is perhaps best known as an astronomer. What matters for us here is that he enjoyed the patronage of the dynasty that was renewing the Muslim expansion into India, and this brought him into contact with Indian culture. Despite the context of his access to India, he did not write as a missionary seeking to deliver the Indians from the darkness of idolatry and bring them to the light of Islam; still less was he a convert to an Indian religion, like the Chinese monks who in earlier centuries came to India in search of Buddhist scriptures and left us accounts of their travels. His concern was quite simply to give an objective account of the Indians. In order to do this, he did something that so far as we know no previous ethnographer, Greek or other, had ever done: he learned the literary language of the people he was writing about in order to read their books. So we could aptly describe Bīrūnī as the world's first orientalist.

Orientalists study people with literary heritages. They are not interested in the Eskimos or the Aranda, nor do they care much for popular culture. In this sense orientalists are elitists, just like Bīrūnī and the judge of Toledo. The reason Bīrūnī thought it worth studying Indian culture was that the Indians were one of the scientific nations. Indeed, he made a point in his book of comparing their ideas with those of the Greeks. It is clear that he would have seen little sense in devoting years of his life to the study of a nonscientific people like the Germans; and the painstaking ethnographies of primitive peoples written by modern anthropologists would have puzzled him. Equally he had no respect for the superstitions of the Indian masses. In this he was not disparaging the Indians as such; he pointed out that the same crude style of thinking was rampant among the masses of ancient Greece, not to mention the Islamic world. It was the elite cultures of civilized peoples that deserved study; the rest was noise.

Bīrūnī, in short, does not offer himself as an icon of diversity. In fact, he could be quite rude even about the Indian elite; he did not, after all, expect them to read his book, nor did they until it was translated into English in the nineteenth century. But both the grand unified theory of the judge of Toledo and the substantive ethnography of Bīrūnī make a significant, if obvious, point about the culture of the Islamic world: that Muslims think thoughts that have nothing to do with Islam. In real life no religion is all-embracing.

III. The Muslim Calendar

The Islamic world arose out of the formation and expansion of a single state. Without that singular political and military origin, it is hard to imagine that a civilization stretching from ocean to ocean could have maintained not just its cultural unity but also a considerable degree of uniformity. Yet no civilization of such an extent could continue to be ruled by one state; the territory was just too large, and, as we have seen, political fragmentation became the normal condition of the Islamic world. One structure that might have helped to maintain the unity of the civilization in the face of this fragmentation would have been a large, powerful organization resembling the medieval European church: a hierarchy including priests, bishops, and cardinals, with the pope at the top, and occasional councils at critical times. But such organizations are by no means easy to hold together. Their very power is liable to generate conflict and resistance, and the outcome may be divisions more intractable than those that would have been there in the absence of the organization. Popes are an invitation to antipopes or, worse yet, to Protestants. It thus makes sense that Islam, which has maintained its unity over the centuries far more effectively than Christianity, took a different route, and an altogether subtler one. We have already looked at one institution that helped to hold the Islamic world together, the pilgrimage to Mecca. Another was the Muslim calendar.

Two aspects of this calendar are of interest in this connection. One is a way of identifying years so familiar to us that we do not easily appreciate its significance. You pick a year, usually one in which an event you regard as significant took place, and call it year 1; you then label all subsequent years by continuing the count. The system that prevails in the world today is of this kind. It emerged in early medieval Europe, and reckons from an erroneous dating of the birth of Jesus. The Muslim system is of the same kind, but starts counting with the year in which Muhammad left Mecca for Yathrib. Such an era seems the obvious way to identify years, but it is far from the only one: the Athenians labeled them with the names of their annual magistrates, many peoples of eastern Eurasia used the twelve-year animal cycle, and the pre-Islamic Arabs would count from some event in the recent past, which they would discard after a while in favor of a more recent one. But if counting in a fixed era is not the only way of doing things, it is in obvious ways the best. The Muslims, though they got there before the Christians, had no priority in adopting such a system. The ancient world was full of eras. But that is exactly the point—there were far too many of them. The Indians had several eras and frequently omitted to say which one they were using; individual cities in the ancient Mediterranean often had eras of their own. In fact, reducing the chaos of ancient chronology to order was a major achievement of scholarship in early modern Europe. What distinguishes the Muslims in this context is that right from the start they picked a single era and stuck to it. This is a great boon to historians: over a vast extent of time and space, events and documents are dated in terms of one and the same era. Having this common era also played a small, but significant, role in maintaining the coherence of Islamic civilization.

The other relevant aspect of the Muslim calendar is more complex. As was alluded to in chapter 5, the Muslims came up with a unique solution to a problem faced by any earthbound calendar maker. This is the awkward fact that—astronomically speaking—the number of days in a month, the number of months in a year, and the number of days in a year are not whole numbers. The unique aspect of the Muslim calendar was its abandonment of the solar year: a Muslim "year" is simply a block of twelve lunar months. The Muslims also allowed the month to float, so to speak: a new month starts when you see the new moon, just as a new day begins when the sun sets (there are complications, such as small print to take account of bad weather, but we can leave these aside). The result is a straightforwardly empirical calendar.

Like any calendar on earth, the Muslim calendar is a package of advantages and disadvantages. There are two obvious disadvantages. One is that the calendar is decoupled from the seasons. Thus Ramaḍān, the month of fasting, comes about eleven days earlier each solar year, moving all around the seasons twice in a lifetime (which is why the pilgrimage to Mecca can fall in high summer). Even in a modern industrial society the seasons make a difference; in a society in which most people are peasants or pastoralists, and the state lives mainly by taxing them, a calendar out of phase with the seasons is very inconvenient. The solution to this problem was that in practice Muslim societies had two calendars: alongside the Muslim calendar there would also be some secular calendar, varying from region to region, and typically of pre-Islamic origin.

The other disadvantage is that the Muslim calendar is not rigid, since a month may contain either twenty-nine days or thirty, depending on the sighting of the next new moon. You may know that a new month will begin toward the middle of next week, but you do not know whether this will happen on Tuesday or Wednesday. Moreover, the new moon may be sighted on different days in different places, especially if they are far apart in longitude. So at any given time the Islamic world may be a calendric patchwork: it could be the first of Ramaḍān in Mecca but not in Baghdad, in Cairo but not in Jerusalem, in this village but not in that. Under modern conditions this lack of rigidity is intolerable; you cannot run an airline (or, for that matter, plan multiple hijackings) under such conditions. But under premodern conditions, where communications were much slower than they are today, such lack of rigidity really did not matter very much.

What, then, was the advantage of the Muslim calendar? Here we need to note that, historically, Eurasian civilizations tended to go for high-maintenance calendars. Keeping the calendar on track typically meant making and enforcing decisions, thus requiring a combination of expertise and executive authority. But experts tend to disagree, and executive authority has a way of fragmenting. When China was divided between more than one dynasty, as happened in the middle of the seventeenth century, there would be more than one calendar in force in different parts of the country. When the pope implemented an eminently sensible reform of the Christian calendar in 1582, it was adopted throughout Catholic Europe within a couple of years; but getting the Protestants on board took the best part of two centuries. An alternative strategy was to devise a very complex calendar, so carefully contrived that it would need no adjustment in any foreseeable future. But if an automated calendar depends on the correct application of complicated rules, then sooner or later someone will make a mistake. The pre-Islamic Persian calendar had a rule requiring the intercalation of a month every 120 years. When the Persian Empire was overthrown by the Muslims, there was no longer a Persian state to implement the rule. The resulting confusion is nowhere more evident than in a three-way calendric split that afflicts the Parsees of Bombay, a community still faithful to the Zoroastrian religion of pre-Islamic Iran.

The elegance of the Muslim calendar was its immunity to all this. As long as the Muslims kept track of the setting of the sun and the appearance of the new moon, and reckoned the years by counting off the months by the dozen, nothing could go wrong with their calendar. The only role for executive authority was to iron out some of the divergences that arose from earlier and later sightings of the new moon. But neither the divergences nor any official action that might be taken to limit them carried the risk of long-term calendric splits, since the calendar was reset for all Muslims by the sighting of the next new moon. It would be hard to think of a calendar more apt to preserve the unity of a politically fragmented, territorially extended civilization.

CHAPTER 13

THE EUROPEAN EXPANSION

I. What the Europeans Did

Starting Points for Old World Expansions

When we considered the prospects for a shift toward the territorial unification of Eurasia, we found ourselves looking at its northern and southern peripheries. In this chapter, by contrast, our concern is with the expansion of Eurasia into the New World and the Antipodes, and here the interesting phenomena cluster at the eastern and western ends of the Eurasian landmass. We have already encountered most of them in one context or another, but it may be helpful to bring them together here (see map 11).

Before we start, it is worth underlining a couple of things that did not happen. First, with one minor exception, there were no maritime expansions into Eurasia, as opposed to out of it. Second, the peoples of Africa played only a limited role in the process. Carthage, by far the most important sea power ever located in Africa, was a Phoenician colony. The ancient Egyptians sailed the Red Sea, but went no farther. In early modern times many Africans crossed the Atlantic to the New World, but only as a result of

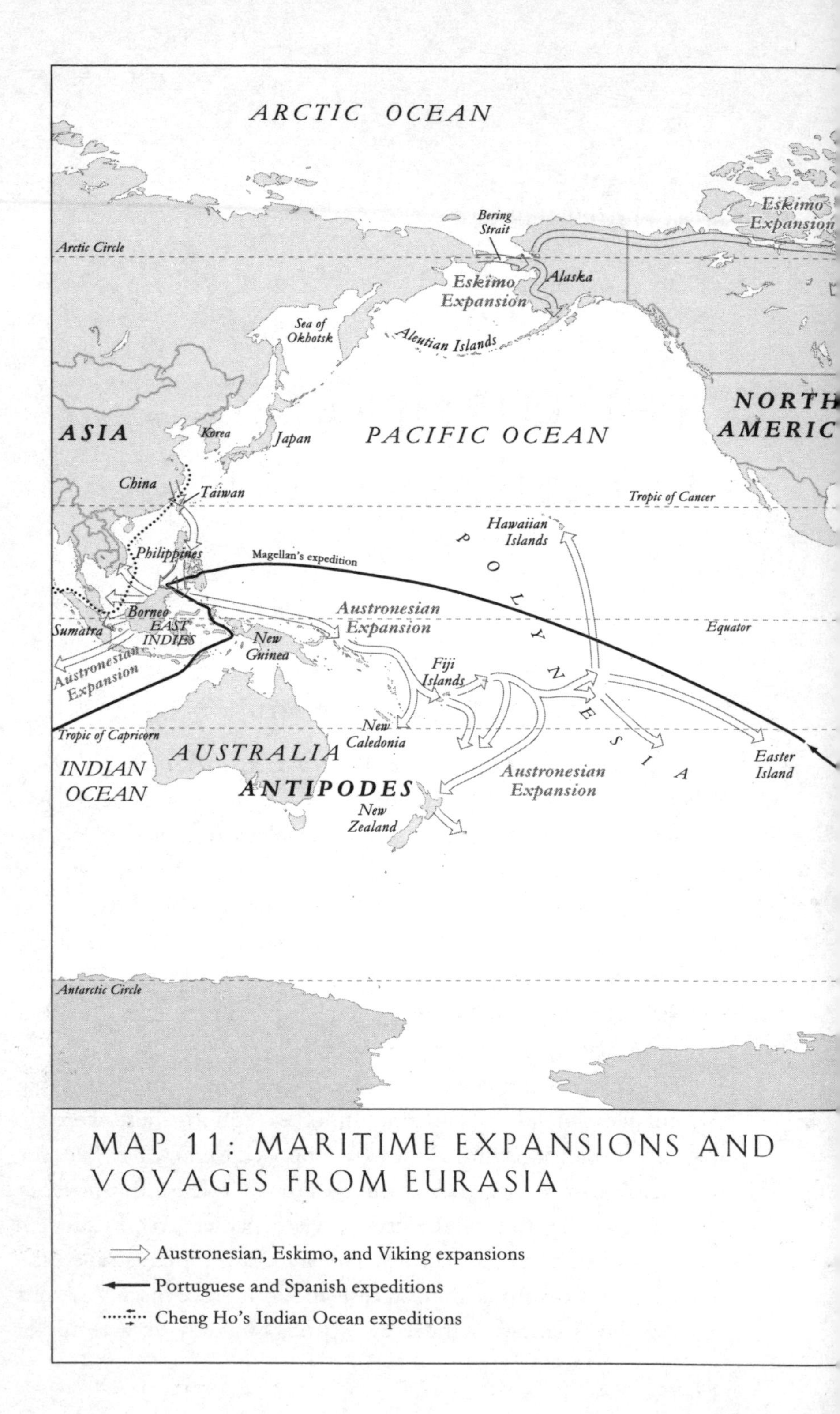

MAP 11: MARITIME EXPANSIONS AND VOYAGES FROM EURASIA

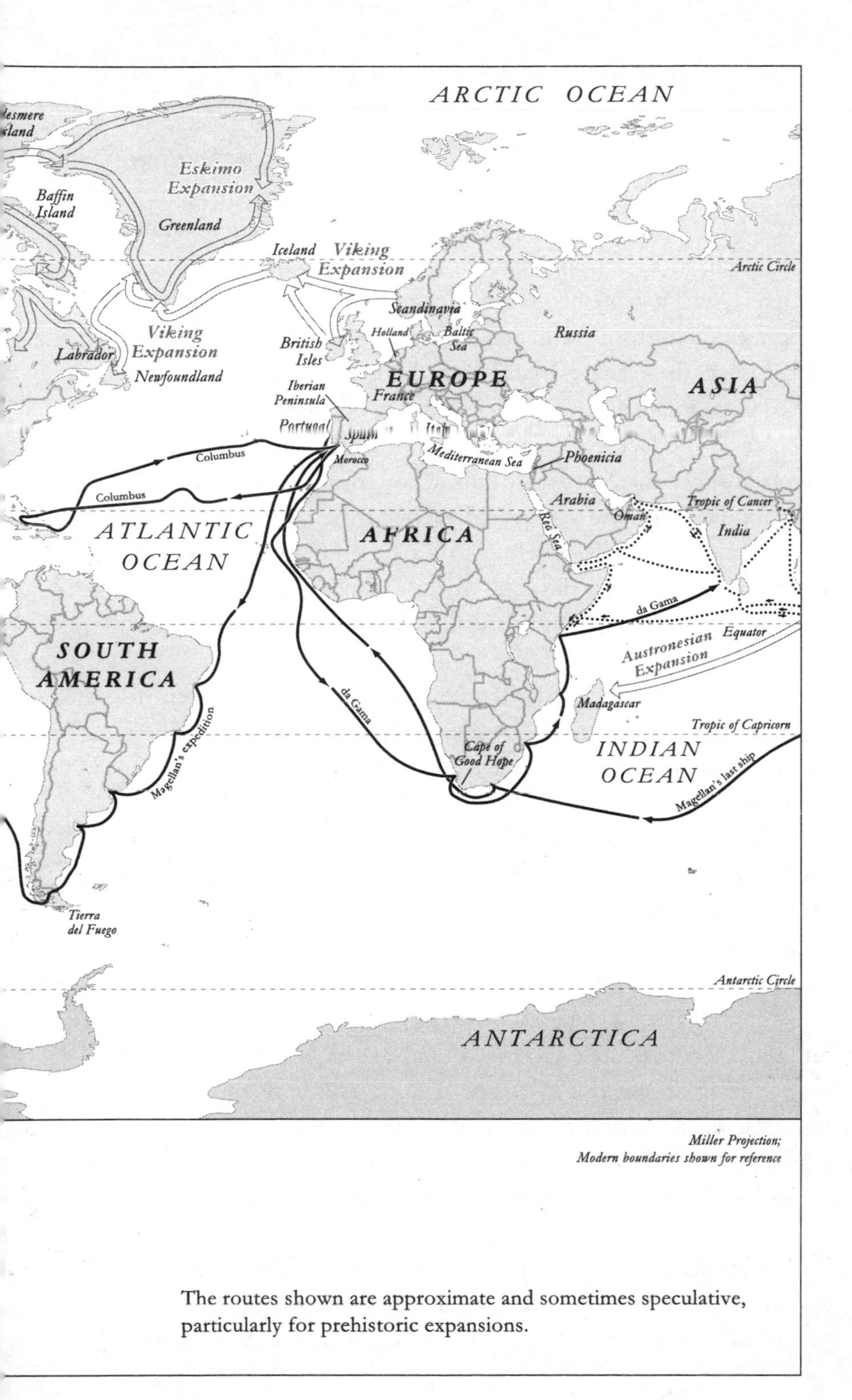

The routes shown are approximate and sometimes speculative, particularly for prehistoric expansions.

being caught up in a slave trade initiated by Europeans. This maritime passivity of Africa is highlighted by the history of some of the islands that lie off its coasts. Madagascar remained uninhabited until colonized by a population from Borneo in the first millennium A.D., and it was only in the course of this settlement that an African element from the mainland appeared on the island. Unlike Madagascar, the Canary Islands off the coast of Morocco were settled from the mainland in ancient times, but contact was lost, as is shown by the fact that Islam did not reach the islands; it was the Europeans, not the North Africans, who rediscovered them in the later Middle Ages and conquered them in the course of the fifteenth century. All this adds up to a pair of familiar points: the advantages enjoyed by the Old World as against the New World and the Antipodes, and within the Old World by Eurasia as against Africa.

The earliest Eurasian maritime expansion may have been the one that took the Eskimos, or their Palaeo-Eskimo forerunners, across the Bering Strait some five thousand years ago. Unlike earlier migrants who had used this route, they did not wait for an ice age; presumably they had canoes comparable to those they still used in recent times, or they may have crossed the sea while it was frozen over in winter. As we saw in chapter 5, the Eskimos possessed a material culture well adapted to the Arctic coastlands of North America, and they occupied this niche all the way across to Greenland. What they did not do on any scale was to move deeper into the Americas. Such a move would have required considerable change to their specialized Arctic culture; and it seems that the mere presence of hunter-gatherers to the south was enough to inhibit such an adaptation, or to limit its success. Thus the Eskimo expansion from northeast Asia was confined to the American Arctic.

The next significant expansion started from the southeast and was much more dramatic. Taking our cue from the family of languages with which it is associated, we can call it the Austronesian expansion. Like the Eskimo migration, it was the work of a prehistoric Stone Age population, and we have to reconstruct it from a combination of linguistic, archaeological, and—more recently—genetic testimony. This evidence combines to reveal a process of eastward migration from island Southeast Asia into the Pacific starting in the second millennium B.C. and reaching Easter Island at the eastern limit of Polynesia around A.D. 500. As we saw in earlier chapters, correlations between linguistic, archaeological, and genetic evidence can be messy. But in this case much of the picture is unusually simple. East of the satellite islands of New Guinea, there had been no earlier human presence, and there is likewise no indication of later migrations from the west on a significant scale. So here the spread of the Austronesian language family, the appearance of the Stone Age Lapita culture or its later derivatives, and the arrival of a particular subset of human genes, can confidently be ascribed to one and the same event. Where the expansion first began is not quite so clear. Linguistic and archaeological evidence would support a homeland in southern China—or, more specifically, in Taiwan, where very diverse Austronesian languages are still spoken among the aboriginal population; but currently the genetic evidence does not sit well with this.

Like the Eskimos, the Austronesians had a material culture notable for its maritime specialization, which in their case was of course tropical rather than Arctic. Moreover, reconstruction of the vocabulary of the protolanguage spoken by the Austronesians outside Taiwan reveals an equivalent of the Indo-European horse and cart: the outrigger canoe (the outrigger is attached to the canoe to keep it from rolling over). With this relatively simple vessel, the Austronesians of island Southeast Asia colonized an environment that has no parallel anywhere else on the globe, the innumerable small islands of the southern Pacific. But they settled few islands of any size; New Zealand and, far to the west, Madagascar were the exceptions. Likewise, they did not reach the Americas, and, a little more surprisingly, they scarcely impinged on Australia. As a result, the impact of the Austronesian expansion was more extensive than intensive.

Another corner of Eurasia that initiated a maritime expansion was the far northwest, as we noted in chapter 5. Unlike the Eskimos and the Austronesians, the Vikings of Scandinavia were an Iron Age people, and their expansion took place at a time when some of their neighbors were sufficiently literate to record their maraudings. The process began in the late eighth century A.D. and was over by the middle of the eleventh. Apart from the British Isles, most of the territory affected lay within the Eurasian landmass, which the Vikings could penetrate effectively, thanks to its rivers. Indeed, their most important historical legacy was perhaps the Russian state. But some Vikings also sailed westward, particularly those who wished to escape the kings who were beginning to dominate their homeland; these colonists (with their Irish slaves) settled in Iceland and, for a while, in Greenland. Farther west a brief Viking presence is well established for Newfoundland in the early eleventh century, but that is as far as it goes. Though the long ships used by the Vikings were certainly more seaworthy than Eskimo or Austronesian canoes, this northern Atlantic expansion left no mark on the American mainland.

At this point there were no more extremities of the Old World from which island-hopping expansions could be mounted. In effect this meant that a Eurasian entry to the New World and the Antipodes would turn on the development of oceangoing vessels in the temperate zone at one end of Eurasia or the other.

At the east end of Eurasia long dugout canoes with some kind of outrigger were in use in Japan by the late Jōmon period, at a time when deep-sea fish already formed part of the Japanese diet. By the sixteenth century A.D. both the Chinese and the Japanese had developed considerable maritime skills, and were using them energetically. In the Chinese case there was extensive seaborne trade in East and Southeast Asia, and this was supplemented in 1405–33 by a series of naval expeditions organized by the Chinese state and led by the eunuch Cheng Ho (Zheng He). While they lasted, these expeditions were large and ranged as far afield as East Africa. In the Japanese case the state played no comparable role, but trade, fishing, and piracy flourished even—or especially—in the absence of a strong central government. Japanese ships visited several destinations in Southeast Asia, and settlements were established there. Nearer home, Japanese pirates had menaced the coasts of Korea and China on and off for centuries. Their depredations climaxed in the 1550s; they had their bases in southern Japan, though many of the supposedly Japanese pirates of this time were in fact Chinese. Given this varied and active maritime scene, it would not have been at all surprising if East Asian shipping had sooner or later found its way to the New World or the Antipodes. But at the point at which the Portuguese appeared in East Asia, in the first half of the sixteenth century, there was no immediate prospect of this.

The maritime scene at the other end of Eurasia was also ancient. The Near Eastern farming package that reached Britain toward 4000 B.C. could have done so only by sea, since Britain had already been cut off from the Continent by rising sea levels some three thousand years earlier. Likewise, in the fourth to second millennia B.C., the communities of the Atlantic fringe constituted something like a single cultural zone linked by the ocean. At the same time the region had the advantage of easy communication with both the Mediterranean and the Baltic; the Mediterranean, as we have seen, was one of the world's most precocious regions of maritime development. The Pacific fringe of East Asia, by contrast, was less closely linked to the Indian Ocean, where interaction was in any case less intense; and access to the Sea of Okhotsk—should you happen to know where it is—did nothing for anyone. The Baltic, Atlantic, and Mediterranean scenes were all very active in the later Middle Ages, and the interaction of the Atlantic and Mediterranean was particularly fertile. It was to be personified in the part played by mariners from the city-states of northern Italy in the Iberian voyages of discovery. But perhaps more important was a confluence of maritime technology that arguably rendered the shipping of the Iberian Peninsula the most likely candidate to breach the traditional limits of Old World seafaring. The key development here was the caravel, which combined the maneuverability of the traditional galley with the relatively small crew of the traditional sailing ship. Yet we should not exaggerate the significance of this technological lead. We have a credible account of a circumnavigation of Africa by the Phoenicians around 600 B.C., and they could doubtless have crossed the Atlantic had they had a mind to.

The Iberian Expansion

Why, then, did the Portuguese and Spanish have a mind to do such things? To recover the main lines of their thinking, we have to think away the New World and at the same time to underline at least two features of the Old. One was the spice trade: the people living at the western end of Eurasia were inordinately fond of expensive culinary commodities that they knew to originate at the eastern end. The other feature was the geopolitical dominance of Islam. Despite the success of the Iberian Christian states in ending Muslim rule in their own backyards, and eventually evicting from the peninsula all Muslims who did not convert to Christianity, they had only chipped away at the western edge of the Muslim world. In the meantime their gains were being dwarfed by the menacing expansion of Muslim rule in the Balkans under the Ottomans. Indeed for Busbecq, a Fleming who was sent on an embassy to Istanbul in the mid-sixteenth century, an Ottoman conquest of the core territories of Europe did not seem far away. Under such conditions, any attempt to mount a massive attack on the Islamic world by land was likely to prove expensive and futile. The question, then, was whether the peninsular peoples could use their maritime resources to outflank their Muslim enemies. Such a strategy might not seem very realistic, but sending a few ships out into the unknown was at least relatively cheap compared with risking an army in the field; and thanks to the Mongol interlude, Europeans now had a better idea what to expect if they succeeded in finding their way to the far side of Islam.

The rest is the story of the Portuguese and Spanish voyages of discovery (see map 11). It began with the gradual exploration of the west coast of Africa by the Portuguese in the middle decades of the fifteenth century. In 1492 it shifted to a different tempo after the rulers of Spain stepped in to finance an improbable scheme touted by Columbus for reaching the East Indies by sailing west—an idea that seemed sound in principle, but in practice relied on worthless geographical data. The Portuguese had rightly turned him down in 1484, and though in 1492 he firmly believed he had succeeded in reaching the East Indies, what he had actually done was, of course, stumble across the New World. The Portuguese then dispatched an expedition led by Vasco da Gama that reached India by way of the Cape of Good Hope in 1498; this route proved viable, and further extensions of it took the Portuguese to Southeast Asia in 1509, China in 1514, and Japan in 1543. The Spanish did not give up easily and in 1519 sent out an expedition under Magellan that reached the East Indies by rounding South America and crossing the Pacific; the route was commercially useless, but the single ship that completed the journey home had the consolation of being the first to circumnavigate the globe. All in all, the themes of the story are men possessed by wild ideas, none wilder than those of Columbus; rulers prepared to make limited investments in these unlikely propositions, especially when competition between states began to raise the stakes; and expeditions that were miserably small by fifteenth-century Chinese standards, but for better or worse achieved far more dramatic results.

As might be expected, the impact of the Iberian expansion on the non-European world varied enormously; we will get more of a sense of this in the next section when we turn to three very different non-European responses. But at this point we need to focus on a basic distinction between the Old World and the New.

In the New World the coming of the Spanish led to a general demographic collapse and the disintegration of the indigenous civilizations. The key factors in this were the susceptibility of New World populations to Old World germs, and the military and technological gap between the native peoples and the invaders. Of the two factors, Spanish germs proved more lethal than Spanish arms; in Mexico even the Tlaxcaltecans suffered massive mortality, despite the fact that they had formed a close alliance with the Spanish (a Tlaxcaltecan had the right to ride a horse and be addressed respectfully as "Don"). The demographic collapse of native America led in turn to the rapid development of a transatlantic slave trade, which brought some ten million Africans to the New World before it ended in the nineteenth century. In the Americas, then, the Europeans could roam more or less at will, introduce new populations, and effectively monopolize state structures.

In the Old World the Iberian impact was quite different. None of the established territorial states went down to the intruders; even the Portuguese intervention in neighboring Morocco was a signal failure. Instead, they sought to dominate the sea, and to that end made efforts to seize islands and promontories that could serve as outposts of empire. The only significant case of early territorial conquest in the Old World was the Spanish adventure in the Philippines, a region of Eurasia so peripheral that in the sixteenth century it was still barely touched by the state structures and high cultures already present in much of island Southeast Asia. Of course there were dreamers. For example, one Spanish Jesuit in 1588 had a plan for a conquest of China in which the Spanish and Portuguese would be aided by Christians from Japan and the Philippines; but this proposal got nowhere.

Joining In

The Iberian expansion did not pass unnoticed in northwestern Europe; if the Spanish could join in, why not others? The English king was already backing expeditions across the Atlantic in the late 1490s, and his French counterpart followed suit in the 1520s. Although their initial efforts did not have much effect, by the seventeenth century the northwestern Europeans—the French and, still more, the English and the Dutch—had taken over much of the commercial and naval role of the Iberians. This was based in part on a new and more effective naval technology and in part on economic advantages. Though the Portuguese retained some of their outposts, naval dominance on the Indian Ocean was largely lost to the newcomers. In the New World the Spanish held on to their territorial empire, as did the Portuguese in Brazil; but the English and Dutch played an increasing role as interlopers, whether as pirates or as traders.

The participation of these new peoples in the European expansion certainly increased its impact on the non-European world. The establishment of colonies in temperate North America and eventually Australia greatly extended the range of European settlement (the former were remarkable not only for their early attainment of independence in 1776–83 but also for their proceeding to constitute themselves a republic). At the same time the non-European parts of the Old World were exposed to empire builders better able, or eventually more disposed, to conquer indigenous territorial states. But the fact is that down to about 1800—which is as far as we need look in this chapter—the only part of the Old World where this had actually happened on a large scale was India. Here in the second half of the eighteenth century, against the background of the collapse of the Moghul hegemony and the rivalries of the European powers, the British laid the foundations of the first western European territorial empire to appear in Asia or Africa. They also set about the first serious study of Indian culture since Bīrūnī (the *Bhagavad Gītā* was translated into English in 1784).

There was, however, an eastern European empire that acquired large Eurasian territories in this period, namely Russia. The first Russian state had emerged in Kiev in the ninth century A.D. through a fusion of a Viking dynasty and a Slav peasantry; it adopted the Orthodox Christianity of the eastern Roman Empire in 989 (the ruler is said to have considered the alternative of converting to Islam, but to have rejected it on the ground that the Russians could not do without their liquor). This state was destroyed by the Mongols in the thirteenth century. But already in the next century a Russian state based in Muscovy was expanding eastward, and by the middle of the seventeenth century it had reached the Pacific. Among other things, this expansion represented a slow but significant triumph of peasants over nomads in northern Eurasia. By the same token, though, the territories accumulated by the Russians prior to the late eighteenth century lay too far to the north for the process to impinge very seriously on the major civilizations of Eurasia; none of these civilizations, after all, had much stake in Siberia. Thus the overland expansion of Russia, unprecedented though it was, did not mark any drastic shift in the balance of power between one Eurasian civilization and another.

All in all, the Eurasian maritime expansion was a clear instance of an accident waiting to happen. Sooner or later, some Eurasian people was bound to intrude into the New World and the Antipodes, with shattering effects on their existing populations. In the event, the intruders were western Europeans; but we should probably not think of this outcome as deeply determined by the course of Eurasian history. Nor should we see the gap between the western Europeans and other Old World populations of this period as a chasm. In the middle of the seventeenth century the Omanis, the Muslim tribal population of southeastern Arabia, happened to capture some European ships at the end of a long struggle against their infidel Portuguese enemies. At their ruler's behest, they used the ships to establish an Omani navy, and went on to engage in commerce, warfare, and colonization in the western Indian Ocean; they did so with a zest that put them at least in the same league as the Portuguese.

If the inhabitants of an arid corner of Arabia could join the maritime club in this fashion, then surely the denizens of more favored lands could have done so too. As we have already seen, the extent of Chinese and Japanese maritime activity in the sixteenth century was considerable. There were many more Chinese than Spanish in Manila when the Chinese rebelled against their Spanish rulers in 1603; there were also a good many Japanese, who played their part in suppressing the Chinese rebellion but themselves threatened revolt in 1606. Indeed, one nightmare scenario entertained by a Spanish governor in Manila in 1605 was that the Japanese would update their navigation and gunnery through contact with the Dutch, and then mount an invasion of the Philippines. This did not happen, though in 1616 one rich Japanese made an abortive attempt to conquer Taiwan. For in contrast to the rulers of Oman, those of China and Japan did not throw the weight of the state behind an East Asian maritime expansion. In 1567 the Chinese state finally abandoned its policy of trying to suppress Chinese participation in overseas trade, and instead began to tolerate it; but it did not go so far as to support it. In Japan feudal anarchy through much of the sixteenth century meant that merchants and pirates could find a variety of local patrons; but the restoration of central government led in due course to the suppression of Japanese maritime activity, not to its encouragement. Although such policies can be explained in terms of the geopolitical structure of East Asia and the traditional agrarian focus of Confucian political culture, the fact remains that at least in the Japanese case the outcome was by no means inevitable. The point is not trivial. Had East Asian mariners received the vigorous backing of their rulers in the early modern period, the world we live in today would be a significantly different place. For a start, the ethnic makeup of the Pacific Rim might be overwhelmingly East Asian (a type of impact the Omanis could not have had in the purely Old World environment of the western Indian Ocean). The plain truth is that a large part of the world was up for grabs. That the western Europeans got so much of it is a distinctly contingent fact.

II. How Some Non-Europeans Responded

What the Europeans did can be told as a single story. But how non-European peoples responded is many stories, and this section will tell only three of them. The first is that of an American people, the Maya; the second is that of an African people, the Congolese; and the third is that of an Asian people, the Japanese. Of course, none of the three can be taken as typical of an entire continent. But their very different stories do illustrate some larger contrasts. As we will see, the Japanese did best in the encounter, the Maya did worst, and the Congolese were somewhere in between; the same is broadly true of Asia, the Americas, and Africa at large.

The Maya

With the arrival of the Spanish the roof fell in on the native polities of Mesoamerica. The lowland Maya of the Yucatán Peninsula were no exception. The first contact came in 1511, and in the following decades Mayan society was devastated by marauding conquistadores, brutally exploitative officials, culturally destructive religious persecution, and, above all, epidemic disease. But if their lot was unenviable, it could have been worse. The chaotic nature of much of the early marauding reflected the Spanish sense that the conquest of Yucatán was a sideshow. The region was too tropical, too politically fragmented, and too run-down to be a major attraction for adventurers in search of wealth and power; in all these respects the Mexican highlands were more inviting. Yucatán thus retained a substantial Mayan population, and the number of Spanish settlers was limited. It is true that—with one exception—there were no longer any independent Mayan rulers, that Mayan society was reshaped by a draconian policy of resettlement, and that the large-scale political, fiscal, and religious institutions of Yucatán were in Spanish hands. But despite all this there was still a coherent Mayan society with its own local elite.

The Maya doubtless had much to say about this experience, but little of it is recorded. There is no surviving Mayan literature from the colonial period written in the traditional hieroglyphic script. This whole cultural tradition disappeared with the conquest and conversion; its last refuge was a remote Mayan kingdom in the Petén that finally succumbed to Spanish conquest in 1697, too early to save modern scholars the labor of decipherment. More surprisingly, there is no significant literature left by acculturated Maya writing in Spanish. One of the more benign features of Spanish rule in the New World was its cross-cultural recognition of aristocracy. Thus in Cuzco, the former capital of the Inca empire, the Inca nobility played a prominent part in colonial society down to the late eighteenth century; their favorite reading included a proud account of the Inca past written in Spanish by Garcilaso de la Vega, the illegitimate son of a Spanish conquistador and an Inca princess. But there was nothing comparable in Yucatán.

What did emerge was something in between: a Mayan literature written in the Latin script. Its characteristic product was a body of texts that are referred to as the books of Chilam Balam. In intention they were works of prophecy, but they provide a good deal of historical information along the way. Books of this kind were produced in various parts of Yucatán, and grew by a process of accretion that lasted into the nineteenth century. Several things show them to be deeply rooted in the indigenous culture of the Maya. "Chilam Balam" means "the spokesman of the jaguar," an animal that had played a part in pre-Columbian religion. The basic framework of the texts was provided by the Mayan calendar, the fundamental assumption being that the events of past calendric cycles could be used to predict those of future ones. The historical material contained in the texts goes back several centuries into the pre-Columbian past, pointing to direct continuity with a hieroglyphic literary tradition. Moreover, we know that the Mayan kingdom in the Petén had prophetic books in the hieroglyphic script. So there is little doubt that the books of Chilam Balam were a continuation of a pre-Columbian Mayan genre.

Yet, whatever the pagan origins of the genre, the texts themselves are not pagan in allegiance. In a rather consistent fashion, they articulate a Mayan view of the conquest that distinguishes sharply between the Spanish and their god. The Maya hated the Spanish, whom they called by a variety of derogatory names, such as "foreign custard apple suckers." The arrival of the aliens, we read, had brought shame and "the beginning of misery"—tribute, tithes, guns, imprisonment for debt, and, above all, forced labor. But with the foreigners came "the fathers of our souls, bringing the word of the True God to your hearts" (the reference is to the friars, who incidentally shared much of the unflattering Mayan view of the conquistadores). There was evident relief that the Christian dispensation meant that in the future "no one will be sacrificed." Moreover, Catholic Christianity had its ways of coming to terms with ethnic particularism; one text refers to Izamal in northern Yucatán as the place "where there descended the daughter of the True God, the Father of Heaven, the Queen, the Virgin"—and indeed the Virgin of Izamal became the patron saint of Yucatán. In addition, the resources of Christianity could be turned against the Spanish. "This," as one text says of the early Spanish depredations, "was the Antichrist here on earth"; God will be angry with the oppressors, and "the justice of our Lord God will descend everywhere in the world."

It is in the nature of prophetic texts that things are seen through a glass darkly. But a more prosaic account of the conquest and its aftermath in the language of a highland Maya group, the Cakchiquel of Guatemala, has a recognizably similar, if more worldly, view of things. The author's approval of Christianity is unqualified. He describes how the friars arrived in 1542 and began instructing the Cakchiquel in their own language; up to that time, he explains, "we did not know the word nor the commandments of God; we had lived in utter darkness." His attitude to the Spanish conquerors is more complex. The Cakchiquel had initially welcomed the intruders as an ally against their enemy the Quiché. Later the relationship turned sour, but our author has no use for a rebel leader who led the Cakchiquel to disaster by claiming that he was the lightning, and would cause the Spanish to perish by fire. Yet he is unmistakably proud of the Cakchiquel resistance to the conquerors, which included some rather effective techniques for making the ground unsafe for their cavalry; and he remembers a year in which "our hearts had some rest" and "we did not submit to the Spanish." Here too, then, the Spanish were one thing and their god was another; our author is neither a pagan nor a collaborator. Of course, this way of looking at things was not universal. Down to the late seventeenth century there were Maya in the Petén who had no more use for the Christian god than they had for the Spanish; at the other extreme, one Mayan chief who wrote his memoirs in northwest Yucatán in the 1560s reveals himself as not only a Christian but also an enthusiastic collaborator.

The accommodation reflected in the persistence of a Mayan prophetic genre in Latin script must have seemed rather stable to many at the time. There was a major Mayan rebellion in 1546–47, and another—the "War of the Castes"—in 1847–55; but between them there was only one notable rising, that of 1761 (significantly, its leader assumed the name of the last ruler of the independent kingdom in the Petén). But from the late eighteenth century on, this accommodation was doomed. Among the immediate causes of this were Spanish administrative reforms that destroyed the autonomy of local Mayan communities, and the massive disruption of native society brought about by the War of the Castes. Looming behind them was the increasing intrusion of the outside world, with its railways, telephones, and markets. When in 1915 the heirs of the rebels found themselves left to their own devices at a time of Mexican civil war, they proceeded to blow up everything that linked them to the outside world.

The Congolese

A people who fared considerably better than the Maya in their early contacts with the Europeans were the Congolese. The mouth of the Congo River inevitably attracted the Portuguese mariners because of the safe anchorage it offered, and from here they made contact with the kingdom of the Congo. Its territory lay in the interior to the south of the river, mostly in what is today Angola. The kingdom, to judge by its own traditions, was established by invaders in the fourteenth century, and it survived in some fashion into the eighteenth. This meant that the relationship between the Congolese and the Portuguese was mediated throughout by an indigenous state. The Portuguese frequently tried to influence the politics of the Congo, with varying success, and in the middle decades of the seventeenth century they made war on it; but they did not conquer it. So whereas the Maya were rapidly overwhelmed, for a long time the Congolese were barely threatened.

In material terms the core of the relationship between the two parties was commercial. Since the kingdom had no previous involvement in long-distance trade, the impact of even relatively minor commercial exchanges was considerable. The Portuguese supplied the Congo with European manufactured goods, especially textiles; soon these were greatly prized among the elite, and the king's ability to hand out such cargo became a major source of royal power and prestige. In return the Portuguese wanted slaves, just like the Muslim traders who frequented the West African interior and the East African coast. But there was a difference: the Portuguese wanted the slaves mainly as an agricultural labor force in the New World, rather than as domestic servants in the Old. This

new Atlantic slave trade offered rich opportunities to some African states, but it was seriously disruptive to others, and still more to stateless peoples. On balance the rulers of the Congo did relatively well out of the trade; their subjects did less well, and some of their neighbors much worse. But in sheer demographic terms, the impact of the slave trade on Africa was far less than that of epidemic disease on the Americas.

The other aspect of the relationship was cultural. Like the Maya, the Congolese adopted Christianity. But the context in which they did so was significantly different, for the Congolese were not simply making the best of it. One reason for this was the contrasting cultural endowments of the two peoples: though the Congolese were able to work iron, they lacked a literate culture. The other reason was the mediating role of the Congolese state, which separated the new religion from the direct exercise of European power. The result was that Christianity had the same attraction for the king of the Congo as it had once had for Ethelbert of Kent. The Congolese king accordingly welcomed the missionaries sent to him by his brother the king of Portugal; he also dispatched Congolese to be educated in Europe (one in due course was made a titular bishop by a reluctant pope). Throughout the process, despite the strong tensions that arose between the king and the Portuguese, the king retained a measure of control.

This comes out clearly in the tensions that developed over the issue of royal polygamy. The Congolese king had many wives, and not just because kings the world over tend to have more than their share of the good things of life. Society was made up of corporate matrilineal groups that played a central role in the political structure of the kingdom. The king's relations with these groups were crucial; he cemented his ties to them by taking wives from them. The attempts of the Christian missionaries to impose monogamy on the royal line thus courted political disaster (here Muslim missionaries would have been more reasonable). On this point the

kings understandably tended to ignore the missionaries; a group of Jesuit zealots who would not leave the issue alone were expelled from the country.

The kingdom of the Congo effectively disintegrated after a disastrous defeat at the hands of the Portuguese in 1665, and though it was able to hang on for a few more decades it has no direct successor at the present day. The state of Angola is the heir of a Portuguese colony, not of an African kingdom. The Congo nevertheless played a precocious role in the history of the modern world. Its relationship with the Portuguese gave rise to the first signs of a phenomenon that was later to become increasingly important, the emergence of Westernized elites in non-Western societies.

The Japanese

If the rulers of the Congo could say no to an impolitic application of Christian morality, those of Japan could say no to Christianity altogether. The Portuguese reached the country in 1543 and brought their usual package of trade and religion. The Japanese proved quite receptive to the new cult, particularly in the south, and the number of converts may have reached 300,000. But Japan, like the Congo, had a state (in fact, as we have seen, a dual one). It so happened that the Portuguese arrived during a prolonged period of civil war and took full advantage of the political fragmentation they encountered; the trick was to convert a local lord, who would then join in forcibly converting his people and destroying Buddhist temples and Shinto shrines in his domain. But in the last decades of the sixteenth century the tide was turning, and early in the next century Japan entered upon the longest period of stability in its history under the Tokugawa shoguns, whose rule was to last well into the nineteenth century. Unlike the rulers of the Congo, the shoguns of seventeenth-century Japan had no need of Christianity, for thanks to the Chinese they already possessed a fully literate culture, indeed a print culture. Instead, they regarded

Christianity, and foreign influence in general, not as a cultural opportunity but rather as a threat to their security (a "pernicious doctrine," as they put it). They therefore decided to close the country: in the future the Japanese would not be let out, and foreigners would not be let in. In 1639 the Portuguese were expelled; thereafter the only Europeans allowed to trade with Japan were the Dutch, and even they had access only to a small island in the port of Nagasaki. Meanwhile, Christianity was so harshly persecuted that by the 1660s it had largely disappeared. A number of things combined to make this closure a viable policy. Japan had a government in firm control of the country; it was remote from the centers of European naval power; and it possessed few natural resources likely to excite the greed of foreigners. The country did, of course, incur serious costs by choosing such a policy, but they were mostly opportunity costs. The interaction that took place under these strange conditions, between the Japanese in their closed country and the Dutch perched in Nagasaki, is actually more intriguing than the original relationship with the Portuguese.

In the process of cultural borrowing, it is often necessary to set aside the charm of an ancestral heritage and adopt something alien but more effective. There are perhaps two fields in which humans are peculiarly sensitive to the pull of what works. At a collective level the consequences of failure on the battlefield make it costly for a society to ignore what foreigners have to teach it about the art of war. And at an individual level something similar is true for the art of medicine. Both these arts have accordingly enjoyed highly cosmopolitan histories. But in Tokugawa Japan the prolonged absence of external threat made it possible to ignore the country's military obsolescence until the rude awakening of 1853, when an American naval squadron under Commodore Perry arrived to demand the reopening of the country. The development that took place from the middle of the seventeenth century onward was accordingly centered on medicine. Even there the limited interest of the state and the restrictions it imposed on contact with the Dutch meant that the interaction was for long rather marginal. Not until the second half of the eighteenth century did a long-standing taste for Dutch exotica (see figure 24 for an example) give way to something more serious, and even then only a relatively small number of people were involved.

A moment of truth, and a very remarkable one, came in 1771 at the dissection of Old Mother Green Tea. The background was a perplexing anatomical problem that had been identified a few years before by two court doctors on the basis of some dissections they had watched. They had carefully compared what they saw with the information given in Chinese medical texts, and had been bothered by the discrepancies. Not unreasonably, they had concluded that there must be physiological differences between Chinese and Japanese. In the meantime two doctors, Sugita Gempaku (d. 1817) and one of his colleagues, had succeeded in obtaining from the Dutch a European treatise on anatomy. They got their chance to put it to the test in 1771 thanks to Old Mother Green Tea, a fifty-year-old woman who had been executed for some awful crime and was then dissected. As usual the work was done by an Eta—a corpse is unclean. The two doctors came armed with their copies of the anatomical treatise and compared what they saw with the illustrations in it. Everything was exactly as shown in the treatise. "The six lobes and two ears of the lungs, and the three lobes on the right and four lobes on the left of the kidneys, such as were always described in the Chinese books of medicine, were not so found." Their conclusion was not that the Chinese differed physiologically from both the Japanese and the Europeans, but that the Chinese had it wrong and the Europeans had it right. So they set about translating the book into Japanese and published it with official approval in 1774. This started a wave of interest in Dutch studies. As early as 1783 one commentator pronounced Chinese learning dead; the Japanese, he urged, should study Dutch learning, which was based on actual facts and not on empty theories. In 1789 the first Dutch language school was opened. In 1815,

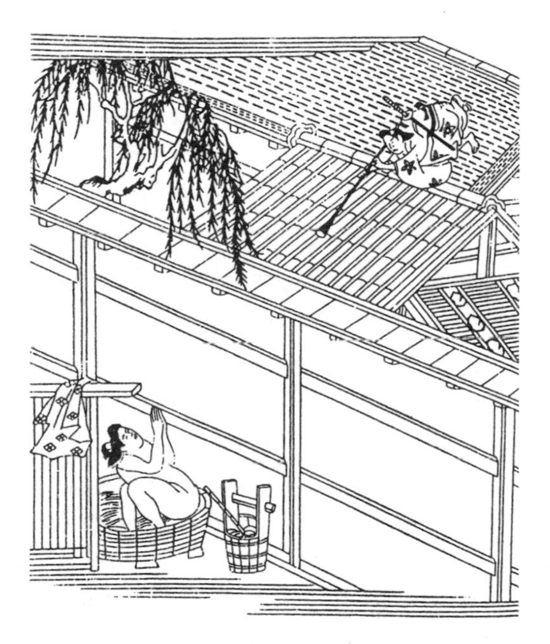

Fig. 24: An early use of a European invention as shown in the Japanese novel *The Man Who Spent His Life in Love* (1682).

near the end of a long life, Gempaku could write, with some exaggeration, "In the beginning there were just the three of us . . . who came together to make plans for our studies. Now, when close to fifty years have elapsed, this learning has reached out to every corner of the land, and each year new translations seem to be brought out."

It is hard not to be impressed by the uncanny precision with which these Japanese physicians homed in on one of the key aspects of European culture, the scientific revolution—and this under historical conditions in which everything European could still be blithely ignored in Japan. That they could be so insightful doubtless reflected the general sophistication of East Asian culture in this period, but it also arose from a more specific development.

In the second half of the seventeenth century, some Japanese Confucians like Itō Jinsai (d. 1705) began to develop an intellectual style profoundly antithetical to the grand Neo-Confucian synthesis that had prevailed in East Asia for so long. The idea was to strip away the inflated metaphysics and distortions of Chu Hsi and his likes and to return to the "ancient interpretation"; this meant restoring the unpretentious straightforwardness that characterized the original Confucian heritage and had been lost for a thousand years. We can see this trend, which had its parallels in China, as a kind of fundamentalism, but it was tempered by Jinsai's belief—shocking to Neo-Confucians—that even a sage like Confucius was not immune to error. Indeed, Jinsai took the view that there are no books in the world that are entirely right—and none, for that matter, that are entirely wrong. The style of this new trend of thought was abrasive. In particular, it was marked by a sweeping contempt for theory and a healthy respect for facts; "a theory," Ogyū Sorai (d. 1728) remarked, "is just a lot of talk."

Despite their rejection of Neo-Confucianism, these scholars were still, of course, fixated on China. Before he turned against Neo-Confucianism, Sorai had buried his first wife according to Chu Hsi's manual of family rituals, an unusual thing to do in Japan, and he became a great admirer of the sixteenth-century Chinese classicists. Later he thought of himself as an "eastern barbarian" who by the grace of heaven had been privileged to revive the authentic Chinese way. Holland for Sorai was no more than a remote country whose language sounded like the chirping of birds or the howling of beasts. Nor was he interested in science: "Not only wind, clouds, lightning, and rain, but the whole movement of heaven and earth are beyond human cognition." To this extent Sorai, to echo an ancient Chinese phrase he was fond of, remained something of a frog in a well. But, as we have seen, other Japanese were now beginning to turn the intellectual style developed by Jinsai and Sorai against Chinese culture at large.

In this they were aided and abetted by another loose fundamentalism of eighteenth-century Japan, a movement that sought to return to the authentic Japanese heritage and liberate it from the overlay of alien Chinese culture. These National Learning scholars were in some ways very different in style from Jinsai and Sorai; for example, they were the champions of a romantic sensibility unknown to formal Confucianism of any vintage. But they shared the disrespect for theory of the Confucian fundamentalists. In the words of Kamo no Mabuchi (d. 1769—the "no" in his name was an affectation with an aroma of Japanese antiquity), "Things which are explained in terms of theories are as dead. Those which operate together with heaven and earth spontaneously are alive and active." Such scholars were happy to use Dutch learning as a stick with which to beat Sinocentric Confucians. One of them went so far as to welcome heliocentrism as a glorification of the sun goddess, suggesting that the idea might have originated in Japan and thence spread to Europe.

There is another point worth emphasizing about the relationship between the Japanese and their traditional Chinese culture. With the passage of the centuries they had come to be very much at home with this culture, but they never forgot its foreign origin. A seventeenth-century Japanese Neo-Confucian put this question to his pupils: "In case China came to attack our country, with Confucius as general and Mencius as lieutenant-general . . . , what do you think we students of Confucius and Mencius ought to do?" The answer was that one should fight and capture the sages alive in fulfillment of one's duty to one's country. In other words, the Japanese, unlike the Chinese, could make a distinction between their identity and their culture, and this encouraged them in a certain eclecticism. Long ago they had adopted what they liked to call the "strong points" of China. If eventually the Japanese came to be disillusioned with the strong points of China, they could substitute the strong points of Europe without disrespect to their ancestors. Indeed, some Japanese of this period went so far as to imagine that the future might outdo the past, an idea that is fairly unusual in traditional societies. An example is the National Learning scholar Motoori Norinaga (d. 1801), who observed, "There are a multitude of things that did not exist in antiquity but are available today, and of things that were inferior in quality in antiquity but are now of superb quality. Judging from this, how can we say that there will not in the future be anything superior to what we have today?"

The early interactions of the Maya, the Congolese, and the Japanese with the Spanish, the Portuguese, and the Dutch were thus very different. But they were by no means arbitrarily different, and they have considerable relevance to the very different histories of these peoples in the modern world.

III. The Sniffing Habit

When the miscalculations of Columbus brought the Old World and the New into collision, the peoples of the New World had the worst of it, and so for the most part did its nonhuman species. But not all American life-forms did badly as the world contracted. There are strong indications that syphilis was exported by the New World to the Old, a small but potent revenge for the infectious diseases that moved in the opposite direction. At the same time a limited number of domesticated plants from the Americas proved very successful on the other side of the world. The Irish became so dependent on the potato that when the crop failed in the 1840s they died in large numbers. The sweet potato came to play a comparable role in the subsistence agriculture of the remote New Guinea highlands. Maize traveled more rapidly through the Old World than it had done in the New. And then, of course, there was tobacco.

Tobacco seems to have originated in the region of the upper Amazon and to have been domesticated there. It spread about as widely in the pre-Columbian Americas as maize. Like maize, it was brought to the Old World by returning Europeans, and there it was disseminated far and wide. By the first half of the seventeenth century, it was well established in Europe, the Near East, India, and China. In each of these areas it was already the object of hostile attention from rulers; James I of England denounced smoking as a "vile barbarous custom."

Tobacco is a drug, and like other drugs, it was used in ways that blended biology and culture. It was widely accredited as a potent form of medicine; this belief remained strong in the Americas and East Asia, though cold water was poured on it in Europe from the early seventeenth century onward. Tobacco was also found effective as a social lubricant; in North American Indian societies it might be used this way in highly formal contexts (as in smoking the pipe of peace), whereas in the Old World it tended to be associated with informal relaxation (Sorai spent his last years at home sipping sake and puffing tobacco with his students). Delivery systems also differed from culture to culture, but were variants on a limited number of themes. You could smoke the stuff, chew it, eat it, or sniff it, and these options were associated with appropriate technologies. Smokers, for example, came to use pipes or cigarettes irrespective of whether they resided in the Old World or in the New (you did not need paper to make a cigarette—the pre-Columbian Mesoamericans used maize husks). But to keep things simple, let us confine ourselves to sniffing.

Sniffers took their tobacco in the powdered form called snuff, inserting it into their nostrils in small doses. You could, of course, do this with nothing more than your hands and fingers. But the tendency was for sniffing cultures to develop more elaborate paraphernalia. The standard South American equipment consisted of trays on which to place the supply of snuff, and tubes through which to sniff it (see figure 25). Such artifacts were used over large parts of the continent from ancient to recent times. Sniffing spread to the Old World with tobacco, but Eurasian cultures did not adopt South American trays and tubes. Instead, they developed small containers in which you could carry your personal

Fig. 25: Sniffing gear from South America: A snuff tube made of bird- and fox-bone, and a whale-bone snuff tray, both from the Chicama valley in coastal Peru (second millennium B.C.).

snuff supply around with you. The Europeans opted for little boxes of various kinds; they were produced in large numbers, and often exquisitely decorated (see figure 26, on the left). But such boxes would not have worked for the Chinese, since they did not do enough to protect the snuff from humidity. Moreover, storing medicine in small bottles was already a long-established practice in China. So while the Europeans favored the snuffbox, the Chinese

Fig. 26: Sniffing gear from Eurasia. *Left:* a French porcelain snuffbox decorated with chinoiserie (about 1730). *Right:* an enameled Chinese snuff bottle decorated with a European woman and child from the reign of the Ch'ien-lung (or Qianlong) emperor (1735–96).

preferred the snuff bottle (see figure 26, on the right). To transfer the snuff from the bottle to the nostril, the Chinese used a small spoon attached to the bottle stopper; Europeans too made occasional use of diminutive spoons, but were normally content to take "a pinch of snuff" between two fingers.

All this provides a nice illustration of the balance that had been struck in the early modern period between the global and the regional. The rapid spread of tobacco—along with the habit of sniffing it—shows how isolation had given way to the beginnings of a more interactive world. Yet this new integration was still a long way from erasing cultural differences. The South Americans continued to place their snuff on trays, the Europeans put it in boxes, and the Chinese in bottles; and each culture for the most part decorated its chosen object according to its inherited artistic canons. There were nevertheless occasional signs that very distinct cultures were beginning to leach into each other. The two eighteenth-century snuff containers in figure 26 have a curious feature in common. The snuffbox on the left is a European product, but it is made of porcelain, an ancient Chinese invention that Europeans had only recently learned to imitate; and the decoration on its lid is an example of "chinoiserie," a faddish European imitation of Chinese art. It is this Chinese look that renders the box exotic, and so doubtless a talking point in the polite European society in which its owner moved. Meanwhile, the bottle on the right, though made in China, uses a technique of enameling recently acquired from the Europeans and depicts a woman and child in a manner that clearly derives from European models. Here, then, it is the European look of the bottle that gives it an exotic cachet.

The symmetry of this reciprocal exoticism is pleasing, but it was not destined to last. The world was set to contract even more.

CHAPTER 14

THE MODERN WORLD

I. Haves and Have-nots

Like the human race itself, many of the really important developments in history have more or less singular origins. The emergence of the modern world in the last two or three centuries is in some ways close to being one of them. If we want to identify the key forces that have shaped the history of the globe over this period, we have to start by concentrating on a single island: Britain. This island was not, of course, a closed country like Japan, and it is unlikely to be an accident that it had been playing an increasingly prominent role in the European expansion overseas, quite apart from its commercial links within Europe. But Britain is where the world we now live in began to take shape.

Britain

The domestic history of Britain over the last millennium has increasingly displayed a number of features that, if seen against the background of the history of the world as a whole, are notably benign. One is the fact that the inhabitants of the island have come to live their lives in conditions of considerable security. Although invasion from the Continent posed an acute danger to the country as late as the Second World War, the last time it actually happened was in 1066. The last civil war was in the mid-seventeenth century, the last major rebellion in 1745. Since then there have been no significant military operations on British soil, though during the Second World War civilians were exposed to sustained bombing from the air. In recent decades there have been race riots but no genocide, small-scale terrorism but as yet no mass murder. Nor do people who live in Britain have very much to fear from lawlessness; in contrast to the way things were as late as the eighteenth century, Britain is an effectively policed society. If we set this experience against the background of widespread and recurrent mayhem that has characterized so much of world history, the degree of security enjoyed by the British has been highly unusual.

Another benign feature of Britain in recent times has been its politics. There has been no break in the legitimacy of government in Britain since the "Glorious Revolution" of 1688. That revolution expelled the last British ruler whose policies raised the specter of absolute monarchy; thus a full-blooded, if syphilitic, king like Henry VIII (1509–47), who put his country through a major religious upheaval to solve his marital problems, has been hard to imagine in British politics for a good three hundred years. Nor have more recent forms of authoritarian rule achieved any success in Britain. There have been no episodes of military dictatorship, no bouts of one-party rule. Instead, forms of limited monarchy and political representation have existed more or less continuously since the Middle Ages. With a series of extensions of the franchise in the later nineteenth and early twentieth centuries, Britain eventually became a democracy—albeit one that still retained some archaic trappings of monarchic and aristocratic power.

A more diffuse accompaniment of this political pattern is the increase in what has come to be called "civility." During the English civil war of the mid-seventeenth century, Englishmen killed each other for God with some abandon. Since then they have stopped doing this. Catholics, long automatically stigmatized as traitors, have been allowed to vote and hold most public offices in Britain since 1829. In Pakistan an incident took place in October 2001 in which three gunmen, doubtless Sunnī Muslims, killed sixteen Protestant Christians in a church; the only thing that was unusual in the Pakistani context was the targeting of Christians, since previously the victims of such attacks had been Shīʿite Muslims. In Britain, by contrast, no one shows much interest in killing people because of their religious beliefs. Ethnicity is more disruptive: there are race riots, and occasional murders inspired by lower-class hatred of immigrants. But the numbers who get killed as a result of ethnic tensions are again minuscule by South Asian standards; although the Welsh, the Scots, and the English do not always see eye to eye, they are not a lethal mixture. Overall, agreement to disagree is pretty much ingrained in the mass of the British population.

These three features of recent British life have their roots in a variety of aspects of the British past, but they can hardly be seen as its inevitable culmination. For example, the fact that Britain has been an island since the early Holocene has clearly worked to its advantage. It helps to explain the relative immunity of the country to foreign invasion; Japan, which is farther removed from the Eurasian coast, has on average done even better over the last two or three thousand years. Yet the dearth of successful foreign invasions has also been a matter of luck—something the world is full of, good and bad. It was the weather that averted a Spanish invasion in 1588, just as it played a major role in defeating the Mongol invasions of Japan. In the same vein we can argue that the continued limitation of the English monarchy in early modern times had something to do with insularity: the lack of acute military threats from outside the island made possible a precocious demilitarization of English society within it. But possibilities do not have to be realized. And even in the unusually civilian environment of sixteenth-century England, the failure of royal absolutism was touch and go: the windfall of ecclesiastical wealth that Henry VIII came into as a result of his Reformation could have rendered the monarchy fiscally independent of Parliament, had the king not proceeded to squander it on the luxury of hostilities with France. So there was nothing foreordained about the trajectory taken by the modern history of Britain. Yet there is one thing that has undoubtedly played a central role in giving the whole syndrome a certain robustness (just how robust only time will tell). This is an economic transformation that started around 1760 and, in a sense, has never stopped: industrialization.

A number of factors were clearly of great importance in making the industrial revolution possible in Britain. They include a rich and varied suite of natural resources, including large amounts of coal; a state that protected the economy more than it crushed it, and supported British trade and settlement overseas; and the replacement of the traditional peasantry by a class of agricultural wage laborers. But given this favorable environment, what actually drove the process was technological innovation in the hands of entrepreneurs. For example, new types of spinning machine drastically reduced the cost of spun cotton, leading to a large expansion of the market for it, which in turn encouraged entrepreneurs to exploit the new technology by reorganizing production in a factory system. The result of such sequences, and of their interactions, was an unprecedented growth of industrial production, at least in certain sectors of the economy and in certain regions of the country. This development was associated with a high level of investment in the transport system (first canals and later railways). Significantly, this too was the work of entrepreneurs.

By 1850 the industrialization of Britain, though still patchy, had made the country far wealthier than it had ever been before. The distribution of the new wealth was markedly unequal, but a wide sector of the population was by now experiencing rising real incomes. Meanwhile, the state had begun to reinvest some of its share of the new wealth in the welfare of its citizens, educating and policing them. By this time Catholics had been emancipated for a generation, and within the next generation a conservative government was to enact the first of a series of measures that led to universal suffrage. Eventually this process of emancipation reached women. With this the basic elements of the British model were in place, and the British began to regard a measure of security, democracy, and civility as their birthright. It would be hard to imagine such a development without the firm foundation of a viable industrial economy.

I have sketched this unprecedented evolution in the form of a story with a measure of explanation worked in. But as with any historical event that is both unique and complex, the explanatory problems are formidable; and in this case they are pressing enough that the ground has been thoroughly trampled by historians. Their competing theories are often plausible, but rarely entirely convincing. Obviously the British transformation was within the range of human possibility; but was it inevitable or even likely that sooner or later a change of this kind should occur? If it was to occur, why not sooner or, for that matter, later? Should we be thinking primarily in terms of a centuries-old British (and more broadly European) divergence from the broader pattern of Eurasian history, or was it more like a sudden lurch into idiosyncrasy in the later eighteenth century? And why did it have to be Britain? Why not the Yangtze delta, for example? As the questions get more specific, the answers can at least become more concrete. Thus we can say with some confidence that the Yangtze delta lacked good access to coal supplies and was ruled by a state with no interest in overseas expansion. But the deeper questions may be too intractable for this kind of treatment. Let us therefore take it as given that the transformation occurred, and prudently move on.

The Rest of the World

So what did this transformation of an island mean for the rest of the world? The results were profound and in many ways disruptive. To put it schematically, there were three basic options for societies outside Britain: to try to carry on as if nothing had happened; to try to adopt some version of the British model, directly or at a remove; or to try to come up with a viable alternative. Much of the history of the world in the last two centuries has been about the choices societies have tried to make in an often unforgiving environment, and the ways in which these choices have played out. Basically, as we will see, the first and last options have not worked; the second has worked in some cases but not in others.

Continuing as if nothing had happened, as the Maya tried to do, was not an inherently foolish strategy. Overreaction is a common human failing, and many of the problems that bother us end up by going away. But in the event no attempt to ignore the modern world proved sustainable, and many ended in disaster. Korea, for example, sought to maintain a closure comparable to that of Japan in the crucial period when the Japanese had reopened their country; until 1872 Korea was ruled by a man who regarded China as the center of civilization and held Europeans and Japanese in contempt. The outcome was that the Japanese retained their independence, while the Koreans succumbed to Japanese rule in 1905; they escaped this predicament only thanks to the American defeat of Japan in the Second World War.

As this example shows, adopting modernity was a better bet than ignoring it, and for many people this strategy worked remarkably well. Modernity, after all, is just a matter of culture, and once the culture has emerged in one place it can be imitated elsewhere; the model is there in front of you. We can divide those who successfully adopted modernity into three main groups. The first is represented by the British overseas. In North America and the Antipodes, British settlers either formed societies in which they constituted the great majority of the population, or established their culture in such a fashion that other European immigrants assimilated to it. These societies adopted the industrial revolution and replicated the other basic aspects of the British pattern with little difficulty; witness the fact that the United States, the world's first modern republic, has for some time possessed the world's largest industrial economy, and wields commensurate power. The second group is the continental Europeans. As early as 1815 some regions of Europe were undergoing the industrial revolution. By the end of the nineteenth century large parts of western and central Europe were firmly anchored in the industrial world, and today the boundaries of successful modernity extend considerably farther. The third group is the East Asians, or some of them. Japan was remaking itself as a modern society in the later nineteenth century, and it remains one today, despite its economic doldrums. The second half of the twentieth century saw similar success among the Chinese and Koreans, wherever political conditions favored free enterprise. Though recent, this development clearly had old roots; Chinese barbers in Mexico City were already competing so successfully in 1635 that their Spanish counterparts sought the intervention of the town council to constrain the market against them. The result of this East Asian modernization is that what started as the British model and later became the Western model has by now achieved a cosmopolitan status.

Industrialization apart, however, the experience of many of these societies was by no means as benign as that of the British. The United States was the scene of a vicious civil war in the 1860s that arose over an institution of very dubious modernity, the plantation slavery to which the country owes its substantial black population. The history of France since the late eighteenth century has been marked by revolutions, invasions, and deep political fault lines—though it is some comfort that the first of these revolutions, that of 1789, played a key role in establishing the modern republican idiom in which even the politics of monarchies are mostly conducted today. Germany, put together as a country only in 1871, lost two world wars and was occupied and partitioned at the end of the second; it was an autocracy till the First World War, a failed democracy between the wars, and a genocidal power in the Second World War. Japan too took an authoritarian path, though a more chaotic one, and likewise suffered defeat and occupation. Nevertheless, the experience of the last half century suggests that by now all these modern societies have acquired the benign features of what began as the British pattern: an industrial economy that serves as the foundation for a fair approximation to security, democracy, and civility for most of the population within their borders. Thus, despite the fact that the world today remains full of wars and rumors of wars, none of them are between successfully modernized countries. Rightly or wrongly, no one worries that a world war will break out because of tensions between the French and the Germans—as happened twice in the last century—or between the Americans and the Japanese, or the Americans and the Europeans.

This may be good news, but it leaves plenty of room for bad news. There remain large parts of the world that have not, or not yet, modernized successfully. Many, of course, have willingly or unwillingly contributed their raw materials and labor to the modernization of others, and provided markets for their products, but for a variety of reasons they themselves have lagged behind. Naturally there is great diversity within this category in terms of both performance and prospects. As of the beginning of the twenty-first century, it is not considered unrealistic to think that China and perhaps India have a chance of success, that the frontier of modernity in Eastern Europe will eventually move a good deal farther east, and that some parts of Latin America (though probably not those inhabited by the Maya) will do better than in the past. On the other hand, there is little optimism to be overheard about sub-Saharan Africa (with the possible exception of South Africa, though not of Angola). More surprisingly, the same is true about the Islamic world, where so far really significant wealth seems to have accrued only to populations with vast amounts of oil (as in Saudi Arabia) or large numbers of Chinese (as in Malaysia). But however grim or rosy the future, as of the beginning of the new millennium the world is full of populations that have yet to derive wealth, security, democracy, or civility from the industrial revolution.

Irrespective of the extent of their success in adopting modernity, all societies that made the attempt faced a problem of which the British (and overseas populations of British origin) were for the most part blissfully ignorant, though Ethelbert of Kent had long ago put his finger on it. In a modernizing world one is liable to discover that—as the Nigerian Christians in the year 2000 remarked uncivilly of Islamic law—one or another aspect of one's inherited culture is "not Y2K compliant." Humans, as we have seen in this book, can be very adaptive, but they do not feel good about abandoning their ancestral cultures in favor of the ways of foreigners. In these circumstances the natural impulse is to compromise, and nationalism is, among other things, the name of this compromise.

If we want an example of nationalism, we need go no farther than Ireland—an island whose historical experience over most of the last few centuries has been altogether less benign than Britain's. It did not take the Irish long to realize that they needed to become part of the modern world brought into existence by their neighbor; indeed, as early as the 1790s, an Irish nationalist was insisting (with vivid use of a high-tech metaphor of the day) that if Ireland were free and well governed, "she would in arts, commerce and manufacture spring up like an air balloon, and leave England behind her at an immense distance." The Irish were also fully aware of the advantages they derived from having become for the most part native speakers of English; one bilingual nationalist observed in 1833 that, in view of "the superior utility of the English tongue, as the medium of all modern communication," he could "witness without a sigh the gradual disuse of Irish." But they had no intention of giving up on being Irish, and to make the point they ended up putting a great deal of energy into a sentimental revival of what had once been their national language, since "a people without a language of its own is only half a nation." Above all, while they might want the British to continue to give them access to their labor market, they had no desire whatever to be ruled by them. "My object," as a radical nationalist remarked in 1847, "is to repeal the conquest—not any part or portion of it but the whole and entire conquest of seven hundred years." In short, the Irish wanted to become modern like the British, but they also wanted to be and rule themselves. Irish nationalism legitimated their attempt to do both. Outside Ireland the precise cultural and political circumstances of such efforts vary from people to people, but the basic idea is everywhere the same: modernize your current culture while holding fast to your ancestral identity.

What this means, ironically, is that nationalists have often done as much to liquidate their traditional cultures as to preserve them; they slaughter sacred cows as blithely as they adopt them, and display little substantive patience for aspects of their cultures that no longer work for them. Of course, drawing the line between preservation and liquidation is a difficult and often contentious matter, as we can show with a handful of Japanese examples from the period after the Meiji Restoration of 1868—the event that represents the end of the dual state and the foundation of modern Japan. Mori Arinori was a secular nationalist who was assassinated in 1889 following a visit to the Ise shrine during which he reputedly used his walking stick to raise the curtain concealing the inner sanctuary; in 1873 he had even proposed that the Japanese language be abandoned in favor of a rationalized form of English. Yet he had great admiration for Kusunoki Masashige, the tragic hero of the fourteenth-century imperial restoration. The no-nonsense rationalist Fukuzawa Yukichi (d. 1901), by contrast, had no use for loyalists who "died like dogs," throwing away their lives in hopeless causes—though he was prudent enough to avoid explicitly using such language in speaking of Masashige, saying rather that "we must admire his spirit, but not take his deeds as our model" (had he been alive today, Fukuzawa argued, Masashige would have acted quite differently). Another of these scandals involved Professor Kume Kunitake, an exact textual scholar in the Chinese tradition that had taken shape in the seventeenth century; in 1891 he published a historical essay on the native Shinto religion of Japan in which he viewed it as "the survival of a primitive form of worship." Predictably the essay was denounced by Shintoists as sacrilegious, and Kume was forced to resign his university position. In the ensuing controversy one intellectual, an anti-nationalist who believed in world government, championed freedom of research; but a progressive nationalist whose peers could read the novels of Walter Scott more easily than the *Tale of Genji* insisted that one must "refrain from making a public problem of anything that relates to the imperial household." The bottom line in all this is perhaps that whatever cultural idiosyncrasies nationalists choose to celebrate, they are constrained by the fact that the elites of all nations have to be pretty much alike if they are to function effectively in a global context. Anything else is a prescription for a life as a frog in a well.

So far we have glanced at two of the options open to traditional societies in an increasingly modern world: to try to carry on as if nothing had happened, or to attempt to adopt some version of the British model. That leaves the third: to seek a radical alternative to the brand of modernity initiated by Britain. This too is not an inherently foolish idea. It stands to reason that British modernity should in many ways be an idiosyncratic product of its local history, and even those who have adopted it have manifestly customized it in numerous respects. For example, the British were Christians, but East Asian modernization has shown conclusively that you do not have to be Christian to be modern; as one nineteenth-century Japanese scholar remarked dismissively before his country's modernization had even begun, "Christianity is Buddhism with hair on it." So to take the argument a stage further, might there not be other and quite different ways to live effectively in the modern world, or even to transform it altogether? The British, after all, had at best muddled through to a parochial modernity whose nature they could hardly be expected to grasp until they had attained it; given hindsight, surely the whole phenomenon called out for a profound rethinking that could then be translated into radical political action. The idea is plausible, but to date the experience of such attempts has not been encouraging.

By far the most sustained of these movements was Marxism, a theory of human society, and especially of industrial society, that was devised in the nineteenth century and extensively applied by the Communist parties of the twentieth. Its most stable component was its aversion to the market economy that had hitherto been the foundation of industrial society. As a political praxis it was originally intended to inform the politics of advanced industrial societies, deftly nudging them into a future of classless freedom and plenty through a process of political revolution. In fact, it turned out to work best as a technique for amassing political and economic power in underdeveloped countries, starting with the Russian revolution of October 1917. And amass power it did: the Soviet Union became a superpower, and Communist China a major player on the international scene. What Marxism did not do was deliver to the societies it ruled either freedom or plenty, and toward the end of the century it collapsed. At the beginning of the twenty-first century, in the countries that matter, Communists have either relinquished political power (as in Russia) or retained it only on the basis of renouncing their traditional hostility to the market (as in China).

One of the key strengths of Marxism lay in its commitment to understanding the future of the modern world, and to living in what would eventually be a transformed version of it. Thus the nightmare for the enemies of Marxism, though by the end of the twentieth century no one any longer remembered this, was that it had come up with a form of society more effective than their own. "We are advancing full steam ahead on the path of industrialization," as Stalin declared in 1929, anticipating that his country would soon leave the "esteemed capitalists" and their "civilization" behind; "we shall see which countries may then be 'classified' as backward and which as advanced" (compare the Irish air balloon). Those who clung to the disorganization of electoral democracy and the market economy would be the dinosaurs of social evolution, and the Marxists with their planned economies and ruthlessly centralized power structures would be its mammals. By the same token, one of the key weaknesses of this futuristic doctrine was that it lacked roots in the inherited beliefs and cultures of ordinary people and had no place for them in its doctrinal system. Thus Marxism deemphasized identity—which may be why, once shown not to work, it disappeared almost without trace. And yet, ironically, some of its major successes in coming to power through revolution had been achieved where Marxists contrived to play the nationalist role as defenders of a country against foreign enemies, and did so better than the nationalists themselves—as in China and Vietnam, though not in Russia.

If there has been another attempt to find an alternative way to live in the modern world that merits our attention, it is the phenomenon known variously as Islamism, political Islam, or Islamic fundamentalism (we are not concerned here with the large number of Muslims who have adopted a historically Western modernity pretty much as is). This movement had roots in the Wahhābism of eighteenth-century Arabia, but as a response to the modern world it took shape in the first decades of the twentieth century and became a major factor in the politics of the Islamic world in its last decades. In terms of strength and weakness, it can be seen as a reversal of Marxism. In intention, at least, Islamic fundamentalism was not at all futuristic; its object was to restore the glory of an ancient past that is part of the cultural heritage of Muslims all over the world. It thus resonated with the values of a vast religious community and could give coherence to a great diversity of Muslim grievances, particularly against the West, ranging from military intrusion to cultural pollution; in such contexts it made effective use of the aptness of traditional monotheism for generating intercultural estrangement (as opposed to the mutual understanding that religion is now expected to promote in the West). Thus unlike Marxism, and to a far greater extent than nationalism, Islamic fundamentalism offered to keep faith with the past of its adherents. But if its linkage to the Islamic heritage was clear and strong, its commitment to full-blooded modernity was by the same token somewhat in doubt. At the beginning of the twenty-first century, the nightmare for the enemies of the Islamic fundamentalists was not that Muslim zealots might inherit the earth through possession of a more effective technique for running it but rather that they might wreck it if they succeeded in borrowing the technology to do so. In its confrontation with the numerous economic, social, and political problems of Muslim populations, the main thing Islamic fundamentalism had to offer was faith.

Unlike Marxism, Islamic fundamentalism seems assured of a certain future, though it is hard to tell how long this may last or what it may be. Fundamentalism might, of course, prove too inflexible to be viable as a way of inhabiting (as opposed to disrupting) the modern world—a case of forlornly notching a boat to mark a spot. In that case, one might anticipate that it would eventually fall into disuse, much as Marxism has already done. Alternatively, fundamentalism may in the end prove flexible enough to be compatible with the substance of modernity—the social and cultural infrastructure of modern wealth and power. In this context it is worth remembering the Meiji Restoration: nineteenth-century Japan successfully gate-crashed the modern world under the pretence of returning to a past that had been in abeyance for the best part of a millennium. It is true that Japanese culture was marked by an eclecticism that does not sit well with current levels of zeal among Islamic fundamentalists. But consider the case of Iran, where this zeal has been gradually dissipating since the fundamentalist revolution of 1979. As of the opening years of the twenty-first century, it is not hard to imagine a future in which the revolutionary heritage, without actually being repudiated, might serve mainly to provide the trappings for a modern nation much like any other. If this happens, fundamentalism in the Iranian context will have become a variant form of nationalism. Sunnī fundamentalism may in the end go the same way. There is already an interesting analogy with Marxism in one respect: fundamentalism has come closest to political success as a form of anti-imperialism. Whatever the eventual outcome, it seems unlikely that Islamic fundamentalism will deliver a different way of being modern, as opposed to a different way of reconciling modernity with an ancient allegiance. But let us return to the firmer ground of the present.

The World Today

All in all, the process that began in eighteenth-century Britain has left no one untouched. It has brought about a more homogeneous world than has ever existed before in human times (consider, for example, the cross-cultural uniformity of the nicotine-delivery systems illustrated in figure 27). But it has also led to great inequalities between peoples. Such inequalities are, of course, nothing new in the history of the world; few major encounters of the last two centuries have been quite so unequal as those that resulted from the Spanish intrusion into the New World in the sixteenth century. But the gross disparity of economic, political, and cultural power brought about by the industrial revolution has in the end touched far more people, and it continues to affect the whole shape of the world we live in. It also generates enormous and predictable resentment. Addressing white Americans at their annual celebration of their independence in 1852, Frederick Douglass, an escaped black slave, told them, "The rich inheritance of justice, liberty, prosper-

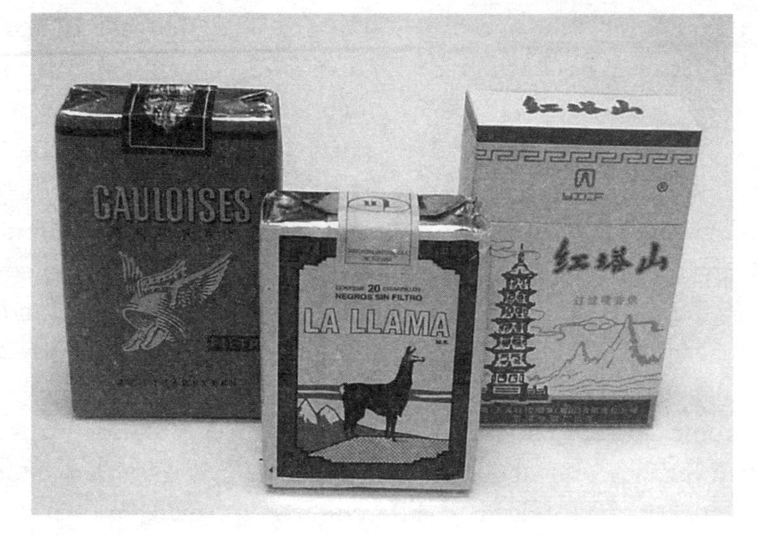

Fig. 27: Nicotine delivery systems of the early twenty-first century from France, South America, and China.

ity, and independence, bequeathed by your fathers, is shared by you, not by me," and ended his oration with the observation that "for revolting barbarity and shameless hypocrisy, America reigns without a rival." The sentiments articulated by Douglass sound very familiar in the world today.

One indication of the degree of this disparity is the almost casual ease with which, in the nineteenth and early twentieth centuries, the British—and others who had successfully joined the imperial club—could rule peoples who remained outside it. The partition of the continent of Africa between the European powers in the late nineteenth century is one example of this, and the creation of a Japanese empire in East Asia at the expense of China and Korea is another. By the middle of the twentieth century the days of such formal empires were largely over. The point was not, however, that modernity had lost its power but rather that it was now differently distributed. The process of modernization had brought into existence modernized elites even within societies that had yet to assimilate modernity on any scale, and the existence of these elites had made direct imperial rule politically and morally hard to sustain, let alone to extend. Hence the world today is for the most part divided into nations rather than empires; even the Russian empire, which thanks to Communism survived much longer than the rest, has now broken up. But behind the façade of the world's united and disunited nations, the formal and informal structures of international relations show that successful modernization remains a precondition for calling the shots.

There is nevertheless one change that has worked to make this unequal world somewhat less callous than it was. In chapter 10 we saw how in the environment of a city-state, as opposed to a territorial kingdom, the elite is obliged to feel some of the pain of the masses. Something similar can be seen on a much larger scale in the world we now live in. In the 1840s it was possible for the British elite to look on with relative indifference as the Irish peasantry starved under British rule a few hundred miles away. Today the British participate in famine relief in regions far more remote, and for which they have no direct political responsibility. One might explain this in terms of a change of sensibility, and without doubt there has been one; the British today care significantly more than they used to about bad things that happen in less privileged parts of the world. But this change of sensibility has received crucial reinforcement from more material considerations. These days awful events unfolding in remote regions have a way of spilling over the borders of the luckless countries in which they take place; they are, for example, liable to generate floods of refugees, and these can be a problem for everyone. The effect, of course, is uneven. A tide of refugees from genocide in central Africa, for example, is unlikely to inundate Europe, whereas refugees from the collapse of states in North Africa or the Balkans are another matter. But the nature of the effect and its relative novelty are clear. The world's haves accordingly display a measure of concern to mitigate the desperation of the have-nots, and more generally of those who find themselves on the wrong side of the current world order. That there was some ground for this concern became evident in the fall of 2001.

II. The Lofty Towers

"Wheresoever ye be, death will overtake you, although ye be in lofty towers" (Koran 4:78). The verse can be applied menacingly to infidels, though in context it is actually addressed to the believers. As often in the Koran, God is encouraging them to be more courageous in fighting his wars, and to this end he is pointing out that since men are going to die anyway, they might as well die gloriously. "Let them therefore fight for the religion of God, who part with the present life in exchange for that which is to come; for whosoever fighteth for the religion of God, whether he be slain, or be victorious, we will surely give him a great reward" (Koran 4:74). It may be a natural human reaction to suppose that it is more rewarding to be victorious than slain, but this is not so: "Thou shalt in no wise reckon those who have been slain in the cause of God, dead; nay, they are sustained alive with their Lord" (Koran 3:169). Better still, could one not be both slain and victorious in the same moment, and thus have the best of both worlds? Whether such a thought can provide an Islamic justification for suicidal attacks mounted by terrorists raises knotty legal questions. But it can certainly generate an intoxicating rhetoric to accompany such attacks.

On September 11, 2001, groups of Muslim terrorists hijacked four airplanes in the United States. They flew two of them into the twin towers of the World Trade Center in New York and one into the Pentagon in Washington, killing several thousand people. The fourth crashed in rural Pennsylvania killing only those aboard it, apparently as a result of the resistance of the passengers. Nobody claimed responsibility for these spectacular attacks, but they were believed with some reason to be the work of an organization known as al-Qāʿida. This was an unusually large terrorist network assembled by the black sheep of a wealthy Saudi family, together with some able associates. The movement had its origins in the successful Islamic resistance to the Soviet occupation of Afghanistan in the 1980s, a resistance that the United States had helped to arm and bankroll through the Pakistanis. After the Soviet withdrawal in 1989, there had been a prolonged Afghan civil war of which the eventual victors were the Taliban, an Islamic fundamentalist militia with close ties to the Pakistanis—and later to al-Qāʿida. The victory of the Taliban, who took Kabul in 1996, gave the network the freedom of much of the country, and it used it to establish and operate terrorist training camps on a scale that dwarfed the activities of such hardy perennials as the Irish Republican Army. The network had already struck against American targets more than once, but the attacks of September 11 were of a different order.

In the last months of 2001 these events riveted everyone's attention, and to some observers they even seemed to have transformed the world. Currents of heady emotion swept through the world of Islam and an unprecedented sense of community developed overnight among countries with a Muslim problem (the sheer extent of the world's Muslim population guaranteeing that there were many such countries). Yet it also made sense to be somewhat skeptical about the long-term effects of the attacks. What had enabled al-Qāʿida to upstage the Irish Republican Army so decisively was two things: its ability to operate in the open in Afghanistan, and the capacity of its operatives to range freely over much of the globe. The sharpness of the international reaction, combined with the scale of the American military response, made it unlikely that either of these conditions would continue to be satisfied. While al-Qāʿida may not have been a spent force, it was hard to see how it could continue to renew itself on the same generous scale. Short of precipitating a biological or nuclear doomsday, it seemed unlikely to be able to shift the large-scale structure of international relations in a direction favorable to its purposes. Indeed, as a player on the international scene it was beginning to look somewhat quixotic. A movement so dedicated to the art of making enemies—one that contrived to antagonize the Americans, the Europeans, the Russians, the Indians, and the Chinese simultaneously—seemed ill adjusted to survival in a predominantly non-Muslim world.

But was the movement really so quixotic? Was slaughtering unbelievers regardless of the consequences really what it was about? Understandably, Americans in particular had a tendency to see it that way, and to take it for granted that the terrorists had targeted Americans out of the depth of their hatred for them. Certainly the organization had no love for them, and no compunction about killing them; it had already declared a holy war against them a few years before. In itself this was not hard to explain. Any nation that happened to be the richest and most powerful in the world would have been more than likely to provoke widespread resentment on the part of those at the bottom of the pile, even if its conduct of its foreign affairs had been a model of tact. Moreover, it was a commonplace in the Muslim world that the United States was hostile to Islam. This was not, of course, an entirely realistic assessment; a superpower has to base its policies on the full diversity of its interests in different parts of the world, and it can ill afford the costs of a consistent global stance for or against anything as ramified as a world religion. But in the Arab world in particular there were specific reasons to see the role of the United States in an unfavorable light—and it was from the Arab countries that the great majority of al-Qāʿida's members seemed to have been recruited.

American policies in the Arab world were shaped around two major commitments, both of them firmly anchored in realities external to it. One was a commitment to the well-being of the State of Israel, and this was locked into the domestic politics of the United States. Israel had emerged during the first half of the twentieth century from an unusual combination of national liberation and colonization; it was the colonization of which the Arabs were more aware, since it had taken place at their expense. Ensuring the well-being of Israel also meant supporting regimes that reached accommodation with it, and were thereby likely to be seen as having betrayed the Arab cause. The other commitment was to the stability of the key regimes in the oil-producing regions of the Persian Gulf, above all Saudi Arabia; this arose from the intractable dependence of the industrialized world, America included, on imported oil. The Saudi state was dominated by a royal family whose reproductive success since the eighteenth century had far outstripped that of any European dynasty; in proportional terms, it was somewhat as if the most desirable opportunities in the United States had been engrossed by half a million descendants of George Washington. This was by any standards a remarkable family achievement, but in the modern world the results were not easy to justify in either a republican or an Islamist idiom. In sum, neither the Israeli nor the Saudi commitment was a way for the United States to look good in an Arab world made up of discontented populations and unpopular regimes. All this guaranteed a vein of virulent anti-Americanism to which al-Qāʿida could effortlessly appeal.

But as one line of commentary on the terrorist attacks pointed out at the time, it did not follow from this that anti-Americanism was the driving force behind al-Qāʿida's ambitions. Its leader's rhetoric was in fact somewhat wooden when he talked about Americans, or even Israelis. Where it came alive was when he placed the Saudi regime between his sights, and it was likely that members of the organization from other parts of the Arab world felt the same way about the regimes in power in their own countries. Their serious strategic goal was most plausibly not simply to vent their feelings or to turn the world upside down but to seize power in Muslim, particularly Arab, countries, and to Islamize them—to

"Talibanize" them, in a phrase that passed into general usage in the aftermath of the attacks.

This ambition was as old as Muslim fundamentalism. For a long time the fundamentalists had tried the obvious tactics: they had sought to infiltrate or overthrow regimes in Muslim countries using whatever means were available to them. By the end of the twentieth century, they had colored the politics of virtually every Muslim country. They had assassinated a president in Egypt; they had achieved, but failed to retain, a guiding role in the government of the Sudan; they had lost a civil war in Algeria, but more or less won one in Afghanistan, where fundamentalists had previously been vouchsafed an unusual opportunity to gain popular support in the course of the resistance to the Soviet occupation. Yet the overall strategy had proved a failure, perhaps because the capacity of the existing regimes to hold on to power was too great, perhaps because fundamentalists mounting attacks in Muslim countries killed fellow Muslims (if only by spraying poison on their prayer beads, as one al-Qāʿida manual recommended)—a tactic guaranteed to divide rather than unite Muslim opinion. Without any question fundamentalism had been most successful in Iran, where the revolution that overthrew the monarchy in 1979 issued in the establishment of an Islamic republic that seemed likely to continue in some form or another for the foreseeable future. But by the end of the century the ruling clerics were losing the moral high ground. The electorate—and Iran was a sufficiently constitutional Islamic republic to have one—seemed increasingly frustrated by the ineffective economic policies of the regime and its cultural repressiveness. And in any case the Shīʿite allegiance of Iran ruled it out as a model for the bulk of the Islamic world. In no other major country had the fundamentalists achieved power by the end of the century. They had acquired extensive cultural clout in the Muslim world and shown a real talent for small-scale social work (though not in Afghanistan), but politically they were going nowhere. In that respect they had signally failed to match the successes of the Communist parties of Eurasia half a century earlier.

The novel strategy of al-Qāʿida may thus have been a piece of lateral thinking. Instead of attacking the regimes themselves, the trick was to attack the Americans. By provoking a massive American retaliation, the terrorists could hope to gain the widespread sympathy of Muslim public opinion and to expose regimes that cooperated with the United States to opprobrium. In the event, however, the resistance to the American invasion of Afghanistan collapsed too quickly for such a process to take hold, and Muslim public opinion remained profoundly ambivalent about the whole affair; witness its tendency to impute the attacks either to the Jews or to the Americans themselves, rather than to glory in them.

It is perfectly possible that in a decade or two these attacks may represent nothing more in the collective memory of Americans than an exceptionally unpleasant episode. But even if the events do not in the long run have much impact on the structures of our world, even if they fail to dislodge the regimes now in power in the Arab countries, they will nevertheless retain a real significance for what they say about the lopsided world we live in. Not least, they remind us once again that there are many people for whom modernity has yet to deliver the benefits that it first bestowed on the British. What distinguishes Muslim populations from others currently on the wrong side of history may not be so much the predicament in which they find themselves, but rather a couple of inherited features that shape their responses: the vast extent of the Muslim community, and the ready availability within its heritage of a particular tradition of religiously sanctioned violence. This tradition may be seized upon by small, fanatical groups whose ultimate purposes have very limited appeal even within the Muslim world, and no intelligibility outside it. But along the way such groups can elicit at least some degree of sympathy from wider populations the sources of whose discontents are readily intelligible to outsiders.

A Latin American parallel can perhaps help to point up the distinctiveness of what has happened in the Muslim world. Take Peru, a country with the usual Latin American resentment of the United States. In recent decades it has suffered from terrorism on a considerable scale at the hands of a Marxist—more specifically Maoist—movement calling itself Shining Path, the objective of which was to overthrow the Peruvian state. It has also experienced the inroads of religious fundamentalism, thanks to the faith and energy of its Pentecostalists. Their main selling point, I was told on a visit to the country, is the social services they provide; in addition, the moral reform they induce in male converts makes the movement attractive to women—sober husbands work harder, bring home more wages, and are less likely to beat up their wives. So Peru too has political terrorism and religious fundamentalism. But in the Peruvian case, as in so many others, the two are quite separate; the terrorists have no religion, and the fundamentalists no politics. What is distinctive about the Muslim world is the extent to which the two elements have come together, and the explosiveness of the mixture.

III. Jupiter's Paramours Again

One central theme in the story of the industrial revolution—what led to it and what it led to—was the accelerating progress of science. Without it, after all, there would have been no 110-story towers and no jet planes for al-Qāʿida to knock them down with. Some of the scientific developments that had taken place by the end of the twentieth century carried the potential to transform human life in ways previously unimaginable; this was the case when scientists began to acquire the rudiments of genetic literacy through their investigations of the workings of DNA. Other advances were simply mind-boggling, like the understanding that had developed of the large-scale history and structure of the universe, and of the behavior of the smallest particles. But one branch of scientific activity was producing results that, though neither obviously useful nor particularly mind-boggling, were undeniably spectacular. This was the exploration of the solar system by unmanned spacecraft.

Like Galileo's research, this exploration depended on state patronage; as Galileo had pointedly remarked, it is rulers who "carry on wars, build and defend fortresses, and in their royal diversions make those great expenditures which neither I nor other private persons may." By the later twentieth century, however, this game was no longer one in which Italian princes or mercantile republics could have competed, even if the last of them had not disappeared in the Italian reunification of the nineteenth century. Despite a measure of international cooperation, there was only one real player, the government of the world's richest country; and even there the politicians blanched at the scale of the costs involved. The government of the United States had nevertheless brought the National Aeronautics and Space Administration (NASA) into existence, and continued to provide it with massive, if unsteady, funding. Here, then, was an organization devoted to the exploration of a domain that prior to the middle of the twentieth century had been utterly inaccessible to humans. NASA in turn employed large numbers of scientists and engineers, where Galileo had lacked even the assistance of a first-year graduate student. The scale of this collective enterprise, like most of its technology, was surely beyond anything he would have imagined.

But many of the fruits of NASA's explorations would not have beggared Galileo's imagination, though his geocentric predecessors with their naked-eye astronomy might have found them hard to assimilate. In fact, several of NASA's findings would have delighted Galileo; like his own discoveries, they would have been grist to his mill. His fundamental point, at which he had hammered away in his discussion of lunar landscapes, was that heavenly bodies were just bodies like any others. This was not, however, a context in which he could derive much mileage from his discovery of the four moons of Jupiter, for even in his largest telescope he could see them only as points of light. Since his day, so many new moons of Jupiter have been discovered that most of us have lost count. But the Galilean four remain by far the largest and most interesting. The successive spacecraft dispatched by NASA in the direction of Jupiter since 1972, and in particular the "Galileo Mission" of the 1990s, have now produced a flood of information about the kinds of bodies these satellites really are.

Io, named after a mistress of Jupiter's who suffered horribly from the jealousy of his wife, is the closest of the four to the planet. For Galileo this meant that it was hard to see (whence the varying number of satellites in figure 22). For us it means that Io is subject to strong tidal forces that tug at its interior, thereby heating it up and generating incessant volcanic activity. This volcanism continually remakes the sulfur-tinged surface of the satellite; where large parts of the surface of our own moon preserve an ancient record of the early history of the solar system, nothing we see on Io is likely to be more than a million years old. Figure 28 (upper left) shows some of Io's mountains as we can now see them; Galileo would immediately have noticed the shadow cast by the setting sun, which just as on our own moon gives an indication of the height of the mountains (in this instance some thirteen thousand feet). Io is, of course, inhospitable to any form of life we can conceive of.

Callisto, named after a princess who became pregnant by Jupiter and had similar problems with his wife, is the outermost of the four. Its surface has none of the youthfulness of Io's. Instead, it is a strikingly early source, marked everywhere by the scars of ancient impact craters (see figure 28, lower right); the later volcanic processes that reshaped significant parts of the surface of our moon have left Callisto untouched. In one respect, however, Callisto is utterly unlike our moon: one of its major ingredients is ice. This raises the interesting possibility that there might be an ocean lurking beneath the satellite's unchanging surface.

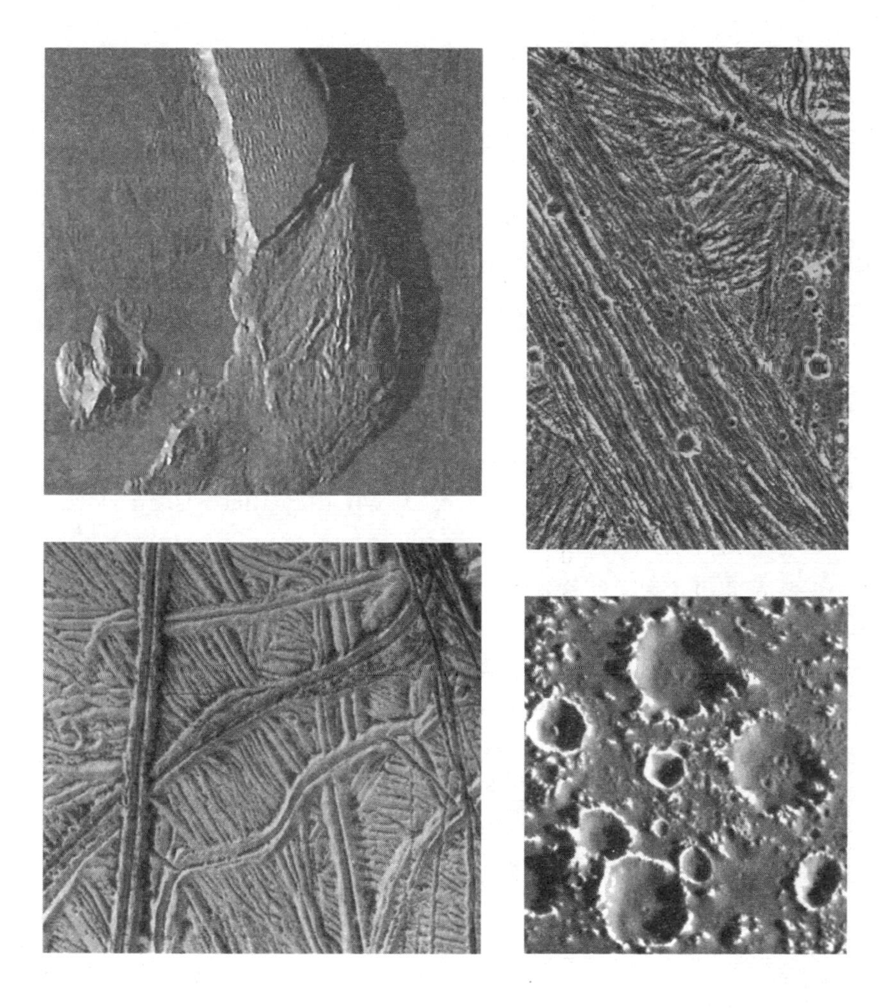

Fig. 28: Late twentieth-century views of the larger Jovian satellites: Io (*upper left*), Europa (*lower left*), Ganymede (*upper right*), and Callisto (*lower right*).

Ganymede, named after a handsome youth kidnapped to serve as Jupiter's cupbearer (and doubtless also as a sex object), is the largest satellite in the solar system, and moves inside the orbit of Callisto. Again, most of what we see is ice scarred by ancient impact craters. But unlike the surface of Callisto, that of Ganymede has been severely affected by later processes, giving the curious impression that some cosmic gardener has worked it over with a rake (see figure 28, upper right). Here there has been no serious talk of a subsurface ocean.

That leaves Europa. Named after a Phoenician princess whom Jupiter abducted to Crete, Europa is the innermost of the Galilean satellites after Io. Here too what we see is mostly ice, worked over so that it looks a bit like the surface of Ganymede, but considerably more bizarre (see figure 28, lower left); a sort of icy volcanism seems to be the main culprit. Just how old this surface may be remains to be determined, but the indications are that it is fairly recent. The question, of course, is whether there is an ocean beneath the ice, and as I write it remains an open question. There could well be more seawater on Europa than here on earth.

All this is very picturesque, and it is a pity that Galileo could not be here to see it. But the single most interesting point is the evidence that Europa, and perhaps not only Europa, may harbor liquid water under its surface. Where there is an ocean there could be life. If there is, that too would have suited Galileo, enabling him to twist the ears of the geocentrists by pointing out yet another respect in which heavenly bodies are not so different from others. But finding life on—or rather in—Europa would also raise a question that Galileo could hardly have posed: How, in biochemical and biological terms, would Europan life compare to our own? This is not, of course, a question about human history, but it is similar in nature to issues of necessity and contingency that we have repeatedly come up against in this book—for example, when we enlisted the Anonymous Conquistador to help us compare the civilizations of the New World with those of the Old. How far are things like life, civilization, and modernization the way they are because that is the only way they could be, if they are to exist at all? And how far is the way they are just one of any number of alternatives? These are interesting questions to puzzle over, but to date our success in answering them, in this book and elsewhere, is limited.

In the meantime the spectacular activities of NASA and al-Qāʿida present a striking contrast. One organization speaks to people who thrive on living in the current global setting, the other to those who hate it. We could see this as a contrast between those who would be frogs around a pool and those who would be frogs in a well. But the two enterprises also have something significant, and very human, in common. Both can serve to remind us of the remarkable extent to which imagination—not to say fantasy—has the power to sideline common sense in human affairs.

CONCLUSION

The account of human history given in this book is broadly materialist. This does not mean that it is based on such manifestly wrongheaded notions as that matter is the only thing in the universe, or that ideas do not matter. It is materialist in the sense that it does not invoke the workings of any supernatural intelligence—divine or other—in order to render the course of history intelligible, or to explain why there has been such a thing at all.

There is nothing new about such an approach. Lucretius, the Roman poet who argued that the universe was not made for us, gave an account of human cultural evolution that was just as materialist in character, and more so in its doctrinal foundations. Though he did not deny the existence of the gods, he located them outside our world altogether and rejected the idea that they in any way concerned themselves with events on earth. What happened in our world (which was just one of many) was in the last resort a matter of the behavior of atoms—"many atoms jumbled in many ways" that "come together and try all combinations." For Lucretius there was thus no teleology to history. Human culture and society were the product of evolution, not of the purposive interventions of gods. At first humans lived like beasts. They lacked such fundamental features of later society as language, agriculture, metal tools, the use of fire, clothes, houses, seafaring, marriage, neighborliness, property, laws, justice for the weak, religion, or writing. Gradually all this changed; clever people introduced innovations, kings founded cities, and so forth. Complex societies emerged, as we would say. This is pretty much the ground I have tried to cover in this book, and in a rather similar spirit—though without the antireligious polemic that is at the core of Lucretius' enterprise.

In some ways we are still pretty much in the same boat as Lucretius when it comes to trying to understand our own history. For one thing, we too have to live with the fact that at least some of human history is yet to be made. Emotionally comforting though this may be, it is intellectually frustrating. For as a vantage point for writing the history of the human race, the present has very little to be said for it, and may well be positively misleading. To an extent we have learned to resist the pull of ethnocentrism, the idea that the most important thing in the world is *who* we happen to be. But we are curiously blind to the self-indulgence of chronocentrism, the idea that the most important thing in the world is *when* we happen to be—that what matters about the past is how it relates to our own transitory present.

The fact is that we, like Lucretius, cannot hope to know the overall shape of human history until it is over—and by then there may be no one left to know it. Perhaps our history will come to an end through easily imaginable acts of human delinquency. It could be the use of weapons of mass destruction, an outcome that might not greatly have surprised Lucretius, who shared with us a vivid sense that progress is not always for the best; in primitive times, he wrote, "thousands of men led by the battle-standard were not wiped out in a single day," as they were in his own times through the invention of "horrible weapons for humanity, increasing day by day the terror of war." Or it could be that the irreversible degradation of the environment will put a stop to our history, an eventuality Lucretius did not anticipate, though he did describe the beginnings of the process when he spoke of metallurgy giving men "tools to fell the forests," and of the way humans "forced the forests to recede up to the mountain, and yield the land below to farming." Or perhaps the end will come through purely natural processes. We do not understand why the Holocene climate has been stable for so long, but the palaeoclimatic record gives us strong indications that this stability has to end sometime—an idea that would not have occurred to Lucretius, though it would certainly have appealed to him, since it aptly illustrates his point that the world was not made for us. Or maybe our history will terminate in some quite different way. And then again, it might continue into a future we cannot even imagine. That might be interesting. But in the end we are not much better informed on such matters than Lucretius was.

Another respect in which we still find ourselves in the same league as Lucretius is that we have nothing that could pass for a scientific understanding of human behavior. Our strongest tool for making sense of our fellow humans, past and present, remains insight, a quality quite distinct from scientific understanding. As the great scientist Charles Darwin wrote of his father, who was not a scientist at all, "The most remarkable power which my father possessed was that of reading the characters, and even the thoughts of those whom he saw even for a short time." The existence of such powers, which most of us share to a lesser degree, is not surprising. The fellow humans we encounter in our daily lives are highly complicated organisms, and it is plausible that we have been selected over a long period for our ability to figure them out. Darwin went on to say that this gift of insight saved his father "from ever making (with one exception, and the character of this man was soon discovered) an unworthy friend." No comparable pressures have operated over thousands of generations to select humans for an ability to discard unworthy scientific hypotheses; indeed, our minds seem on balance to be better adapted to religion, which projects onto nature the intentions we have evolved to discern in our fellow humans. This is not to deny the possibility that a successful science of human behavior may eventually emerge, though a human is vastly more complex than an electron, a glacier, or a star. But as things stand, we are not much closer to it than Lucretius was.

There are, of course, many respects in which we can now do better than Lucretius could possibly have done. We have some understanding of genetics and of the features that human societies share with those of other primates; he had virtually none, and his account suffers from it. He had no archaeological evidence to apply to the reconstruction of prehistory, and no dating techniques. He also knew less than we do about most human societies, including many that existed in his own day (though unlike us he would have had no trouble recognizing the loom weights of figure 1, which are all that survive today of looms like that in figure 19). Naturally he knew nothing of the two thousand–odd years of history that lie between him and us. In ways like this we can validly claim to have left Lucretius behind.

There is, however, one prominent theme of this book that Lucretius certainly could have taken up had he chosen to: human cultural diversity. Though very much aware of the importance of cultural change, he was not much interested in cultural differentiation. Thus he spoke of the spread of the idea of gods to all nations, but had nothing to say of the variety of their religious conceptions. In part this may reflect a major shift in perspective that has taken place between him and us. The unprecedented conditions of the modern world have made the full range of human cultural diversity accessible to study for the first time, with the result that we now know a great deal more about it. But the same conditions have also threatened much of this diversity with extinction, a process that was making a dramatic impact on the Americas as early as the sixteenth century and that today affects even a country like France. The result is that we have come to value diversity because, like farmland in New Jersey and rainforest in Brazil, it is something we are losing. This in turn perhaps inclines us to exaggerate its significance. Walls, pottery, and writing, for example, are immediately recognizable across the cultures that possess them; and to my knowledge no one has yet come up with a feature of one culture that could not in principle be made intelligible in another. But differences there have been in plenty, and there is a better reason than any change in perspective why a book on human history should focus on cultural diversity.

Consider the history of this diversity over the long term. For the most part we can only speculate about it, but it would probably be a mistake to think of it as having been more or less constant until recent times. If modern humans go back to a singular origin in a small interbreeding population, as seems likely to be the case, then the chances are that they initially shared a single culture. Presumably this culture subsequently differentiated as the territory occupied by modern humans became more and more extensive. Perhaps the zenith of cultural diversity was in the later stages of the Upper Palaeolithic, with its subsequent trajectory being downhill all the way; it is hard to tell, since it is only from the point at which literate societies start to leave substantial written records, and illiterate ones begin to be described by the literate, that we can hope to have a serious picture of the range of cultural diversity. Certainly the emergence of civilization must have affected it badly, since at least at the elite level it created relatively homogenized cultures that held sway over wide territories. To this extent what has happened in modern times can be seen as an intensification of a process that was already at work long before.

The fact of the matter is that cultural diversity contains the seeds of its own decay—or of progress, if one prefers to put it that way. It is for this reason that it is so central to an understanding of the course of history. As Lucretius was acutely aware with

1. *The Palaeolithic Background*

There are numerous books on this subject. To name two of them: Richard G. Klein, *The Human Career: Human Biological and Cultural Origins*, 2d ed. (Chicago and London: University of Chicago Press, 1999), a clear and comprehensive textbook with an individual slant; Robert Foley, *Humans before Humanity* (Oxford and Cambridge, Mass.: Blackwell, 1995), a book for the nonspecialist marked by consistently sharp analysis. Climatic change: Richard B. Alley, *The Two-Mile Time Machine: Ice-cores, Abrupt Climate Change, and Our Future* (Princeton: Princeton University Press, 2000); the substance of the book concerns the past. Human genetics: Matt Ridley, *Genome: The Autobiography of a Species in 23 Chapters* (New York: HarperCollins, 1999), a readable introduction; Luigi Luca Cavalli-Sforza, *Genes, Peoples, and Languages* (New York: North Point Press, 2000), a short survey of the implications of genetic studies for the study of the human past by a veteran of the field. Stone tools: Kathy D. Schick and Nicholas Toth, *Making Silent Stones Speak: Human Evolution and the Dawn of Technology* (New York: Simon & Schuster, 1993), describing research on the making and use of stone tools.

2. *The Neolithic Revolution*

The emergence of farming: Bruce D. Smith, *The Emergence of Agriculture* (New York: Scientific American Library, 1995), a broad survey by an Americanist; Jared Diamond, *Guns, Germs, and Steel: The Fates of Human Societies* (New York and London: W. W. Norton, 1997), chaps. 4–10 (updated in his article "Evolution, Consequences and Future of Plant and Animal Domestication," *Nature* 418 [2002]: 700–707). For the Near East: James Mellaart, *The Neolithic of the Near East* (London: Thames and Hudson, 1975), still a valuable survey. Pottery: William K. Barnett and John W. Hoopes, eds., *The Emergence of Pottery: Technology and Innovation in Ancient Societies* (Washington, D.C., and London: Smithsonian Institution Press, 1995), a collection of papers on a wide range of regions.

3. *The Emergence of Civilization*

For the individual civilizations, see below. Writing: Peter T. Daniels and William Bright, *The World's Writing Systems* (New York and Oxford: Oxford University Press, 1996), an accessible reference work for scripts of all periods.

4. *Australia*

Australian prehistory: John Mulvaney and Johan Kamminga, *Prehistory of Australia* (Washington, D.C., and London: Smithsonian Institution Press, 1999), a textbook treatment. The Aranda: Baldwin Spencer and F. J. Gillin, *The Arunta: A Study of a Stone Age People* (London: Macmillan, 1927), a detailed ethnography.

5. *The Americas*

The peoples of the Americas: *The Cambridge History of the Native Peoples of the Americas* (Cambridge: Cambridge University Press, 1996–2000), a multivolume collective work organized by region: North America, Mesoamerica, South America. The American civilizations: Richard E. W. Adams, *Ancient Civilizations of the New World* (Boulder: Westview Press, 1997), a convenient short overview of both regions. Mesoamerican civilization: Michael D. Coe, *Mexico from the Olmecs to the Aztecs*, 4th ed. (New York: Thames and Hudson, 1994). Mesoamerican calendars: Joyce Marcus, *Mesoamerican Writing Systems: Propaganda, Myth, and History in Four Ancient Civilizations* (Princeton: Princeton University Press, 1992), chap. 4.

6. *Africa*

African prehistory: David W. Phillipson, *African Archaeology*, 2d ed. (Cambridge: Cambridge University Press, 1993). The history of Africa: J. D. Fage, with William Tordoff, *A History of Africa*, 4th ed. (London and New York: Routledge, 2002), a one-volume survey; *The*

Cambridge History of Africa (Cambridge: Cambridge University Press, 1975–86), a multivolume collective work organized by period. Ancient Egypt: Ian Shaw, ed., *The Oxford History of Ancient Egypt* (Oxford and New York: Oxford University Press, 2000), a one-volume collective work. The Samburu: Paul Spencer, *The Samburu: A Study of Gerontocracy in a Nomadic Tribe* (London: Routledge and Kegan Paul, 1965), a readable ethnography. Age-group systems wherever they are found: Frank Henderson Stewart, *Fundamentals of Age-group Systems* (New York: Academic Press, 1977), unique in its coverage but technical. My own discussion of the topic draws on the same author's unpublished paper "Age-group systems in East Africa."

7. The Ancient Near East

The emergence of Mesopotamian civilization: J. N. Postgate, *Early Mesopotamia: Society and Economy at the Dawn of History* (London and New York: Routledge, 1992). The history of the Near East down to the Macedonian conquest: Amélie Kuhrt, *The Ancient Near East, c. 3000–330 BC* (London and New York: Routledge, 1995), a comprehensive survey. The emergence of monotheism: Mark S. Smith, *The Origins of Biblical Monotheism: Israel's Polytheistic Background and the Ugaritic Texts* (Oxford: Oxford University Press, 2001), one of many books on the subject.

8. India

The prehistory of India: Bridget and Raymond Allchin, *Origins of a Civilization: The Prehistory and Early Archaeology of South Asia* (New Delhi: Viking, 1997). The history of India: Hermann Kulke and Dietmar Rothermund, *A History of India*, 3d ed. (London and New York: Routledge, 1998). Culture and society: A. L. Basham, *The Wonder That Was India: A Survey of the Culture of the Indian Sub-continent before the Coming of the Muslims*, 3d ed. (New York: Taplinger, 1968), a fine survey of its field despite its title, which was not of the author's choosing. Telugu courtesan poetry: A. K. Ramanujan, Velcheru Narayana Rao, and David Shulman, *When God Is a Customer: Telugu Courtesan Songs by Kṣetrayya and Others* (Berkeley: University of California Press, 1994), with an insightful introduction and attractive translations.

9. China

The prehistory of China: Kwang-chih Chang, *The Archaeology of Ancient China*, 4th ed. (New Haven: Yale University Press, 1986), still a good survey. The history of China: Patricia Buckley Ebrey, *The Cambridge Illustrated History of China* (Cambridge: Cambridge University Press, 1996), an attractive one-volume survey; *The Cambridge History of Ancient China from the Origins of Civilization to 221 B.C.* (Cambridge: Cambridge University Press, 1999), a large collective volume on the early period; F. W. Mote, *Imperial China, 900–1800* (Cambridge, Mass., and London: Harvard University Press, 1999), a substantial one-volume treatment; *The Cambridge History of China* (Cambridge: Cambridge University Press, 1978–), a massive multivolume collective work covering the period from the Han dynasty to modern times, organized by dynasties, and still in progress. The history of Japan: *The Cambridge History of Japan* (Cambridge: Cambridge University Press, 1988–99), a multivolume collective work. The ancestor cult: Patricia Buckley Ebrey, *Chu Hsi's Family Rituals: A Twelfth-Century Chinese Manual for the Performance of Cappings, Weddings, Funerals, and Ancestral Rites* (Princeton: Princeton University Press, 1991), a translation with a helpful introduction. Antiquarians, fakes, and the history of art in several cultures: Joseph Alsop, *The Rare Art Traditions: The History of Art Collecting and Its Linked Phenomena Wherever These Have Appeared* (New York: Harper & Row, 1982).

10. The Ancient Mediterranean World

The Greeks: John Boardman, Jasper Griffin, and Oswyn Murray, *The Oxford History of Greece and the Hellenistic World* (Oxford: Oxford University Press, 1991), a one-volume collective work. The Romans: John

Boardman, Jasper Griffin, and Oswyn Murray, *The Oxford History of the Roman World* (Oxford: Oxford University Press, 1991), a sister volume. Comprehensive and detailed coverage: *The Cambridge Ancient History* (Cambridge: Cambridge University Press), a massive multivolume collective work (look for the most recent edition of the volume that concerns you). Athenian democracy: Aristotle's *Athenian Constitution*, in John Warrington, trans., *Aristotle's Politics and Athenian Constitution* (London and New York: J. M. Dent and E. P. Dutton, 1959). Attic black- and red-figure: R. M. Cook, *Greek Painted Pottery*, 2d ed. (London: Methuen, 1972).

11. Western Europe

A broad survey: Mortimer Chambers, Raymond Grew, David Herlihy, Theodore K. Rabb, and Isser Woloch, *The Western Experience*, 4th ed. (New York: Alfred A. Knopf, 1987), a convenient and well-illustrated introduction, including also ancient Greece and Rome. Medieval Europe: Robert Bartlett, *The Making of Europe: Conquest, Colonization, and Cultural Change, 950–1350* (Princeton: Princeton University Press, 1993), a fine synthesis. Early modern Europe: Perry Anderson, *Lineages of the Absolutist State* (London: NLB, 1974), insightful characterizations of the states of early modern Europe. John Knox: Marvin A. Breslow, ed., *The Political Writings of John Knox: The First Blast of the Trumpet against the Monstrous Regiment of Women and Other Selected Works* (London and Toronto: Associated University Presses, 1985). Galileo: Stillman Drake, trans., *Discoveries and Opinions of Galileo* (New York: Doubleday, 1957).

12. Islamic Civilization

The Indo-Europeans: J. P. Mallory, *In Search of the Indo-Europeans: Language, Archaeology and Myth* (London: Thames and Hudson, 1989), adhering to a traditional but probably correct approach. The Mongols: David Morgan, *The Mongols* (Oxford and New York: Basil Blackwell, 1986), a readable survey. Islamic history: Francis Robinson, ed.,

The Cambridge Illustrated History of the Islamic World (Cambridge: Cambridge University Press, 1996), a good collective introduction; Ira M. Lapidus, *A History of Islamic Societies* (Cambridge: Cambridge University Press, 1988), a comprehensive survey; Bernard Lewis, *The Middle East: A Brief History of the Last 2,000 Years* (New York: Scribner, 1995), a concise survey of the history of the core of the Islamic world. Bīrūnī on India: Edward C. Sachau, trans., *Alberuni's India*, abr. ed. (New York: W. W. Norton, 1971).

13. The European Expansion

World history since late medieval times: Robert Tignor, Jeremy Adelman, Stephen Aron, Stephen Kotkin, Suzanne Marchand, Gyan Prakash, and Michael Tsin, *Worlds Together, Worlds Apart: A History of the Modern World from the Mongol Empire to the Present* (New York and London: W. W. Norton, 2002), a collective work, but well integrated, with useful annotated sections on further reading. The Maya: Nancy M. Farriss, *Maya Society under Colonial Rule: The Collective Enterprise of Survival* (Princeton: Princeton University Press, 1984). The Congolese: Anne Hilton, *The Kingdom of Kongo* (Oxford: Clarendon Press, 1985). The Japanese: Donald Keene, *The Japanese Discovery of Europe, 1720–1830* (Stanford: Stanford University Press, 1969).

14. The Modern World

Worlds Together, Worlds Apart continues to provide broad coverage and sections on further reading. The literature on the events of September 11, 2001, is likely to blossom, but for an early attempt to take stock see James F. Hoge and Gideon Rose, eds., *How Did This Happen? Terrorism and the New War* (New York: Public Affairs, 2001), an uneven collection of essays on various aspects of the events (on the question of motive, see particularly the paper by Michael Doran). The Iranian chapter in the history of political Islam: Roy Mottahedeh, *The Mantle of the Prophet: Religion and Politics in Iran* (New York: Simon & Schuster,

1985), an engagingly humane introduction. NASA and the moons of Jupiter: Daniel Fischer, *Mission Jupiter: The Spectacular Journey of the Galileo Spacecraft* (New York: Springer-Verlag, 2001).

Conclusion

Lucretius, *On the Nature of Things*, trans. Anthony M. Esolen (Baltimore and London: Johns Hopkins University Press, 1995), a vivid translation.

CREDITS

p. iv, 1, 53, 123, 265: James Davis/Axiom.

p. xx: Courtesy of the author.

p. 4: Photograph of excavated high status residential architecture in Sector B of the Caral site, Supe Valley, Peru, July 2000; courtesy of Jonathan Haas.

pp. 6–7: Reprinted by permission from *Nature*, vol 364, p. 218, copyright July 15, 1993, Macmillan Publishers Ltd.

pp. 16–17: Photographs courtesy of Giancarlo Ligabue, Centro Studi Ricerche Ligabue, from Ligabue missions, 1986–1990.

p. 27: (*Top*) courtesy of the Ashmolean Museum, University of Oxford; (*bottom*) drawings by R. Freyman and N. Toth after Mary Leakey.

p. 33: Drawings by G. C. Hillman.

p. 35: Courtesy of the author.

p. 48: Courtesy and by permission of Joyce Marcus, drawing by John Klausmeyer.

p. 49: Courtesy of Jürgen Liepe Photo Archive.

p. 61: Adapted from John Mulvaney and Johan Kamminga, *Prehistory of Australia*, Washington, D.C.: Smithsonian Institution Press, 1999, p. 55.

pp. 70–71: From John Mulvaney and Johan Kamminga, *Prehistory of Australia*, Sydney, Australia: Allen & Unwin, 1999.

p. 77: Adapted from J. D. Jennings (ed.), *Ancient North Americans*, New York: W. H. Freeman, 1983, pp. 32–33 and *The Times Atlas of the World*, Comprehensive Edition, London: Times Books, 1980, plate 5.

p. 95: Transparency no. 3614(2) (photo by Perkins/Beckett), courtesy the Library, American Museum of Natural History.

p. 101: Adapted from W. M. Adams et al. (eds.), *The Physical Geography of Africa*, Oxford: Oxford University Press, 1996, p. 165.

p. 119: Courtesy of the Griffith Institute, Ashmolean Museum, Oxford.

pp. 128–29: Adapted from W. B. Fisher, *The Middle East*, London: Methuen, 1978, p. 4.

p. 131: Drawing after James Mellaart, from his book *The Neolithic of the Near East*, London: Thames & Hudson Ltd., 1975, p. 101.

p. 145: Courtesy of the Arkeoloji Müzeleri, Istanbul.

pp. 150–51: Adapted from Bridget and Raymond Allchin, *Origins of a Civilization: The Prehistory and Early Archaeology of South Asia*, New Delhi: Viking, 1997, p. 17.

p. 178: Inset adapted from T. R. Tregear, *China: A Geographical Survey*, New York: Halstead Press, 1980, p. 22.

pp. 180–81: Reprinted with the permission of Cambridge University Press from Michael Loewe and Edward L. Shaughnessy (eds.), *The Cambridge History of Ancient China from the Origins of Civilization to 221 B.C*, Cambridge: Cambridge University Press, 1999, pp. 148 and xxii.

p. 203: Ting from the *K'ao ku t'u* by Lü Ta-lin, courtesy of the Harvard-Yenching Library.

pp. 208–9: Adapted from L. Jeftic et al., *Climatic Change and the Mediterranean*, London: Edward Arnold, division of Hodder & Stoughton, 1992, p. 3.

pp. 228–29: (*From left to right*) courtesy of the National Archaeological Museum, Athens; Scala/Art Resource, NY; the Pan Painter, *Bell krater (mixing bowl)*, Greek, Early Classical Period, about 470 B.C., Italy, said to be from Cumae, place of manufacture, Greece, Attica, Athens, Ceramic, Red Figure, height 37 cm, diameter 42.5 cm, copyright 2002 Museum of Fine

Arts, Boston, James Fund and by Special Contribution, 10.185; American School of Classical Studies at Athens: Agora Excavations.

p. 230: From A. Pekridou-Gorecki, *Mode im antiken Griechenland*, Munich: C. H. Beck, 1989, p. 43.

p. 232: Courtesy of the Martin von Wagner Museum, Universität Würzburg. Photographs: K. Oehrlein.

p. 233: (*Left*) The Art Museum, Princeton University, gift of Frederick H. Schultz, Jr., Class of 1976, photograph: Bruce M. White; (*right*) Princeton University Art Museum, museum purchase and gift of Nicholas Zoullas, photograph: Bruce M. White.

pp. 236–37: Adapted from N. J. G. Pounds, *An Historical Geography of Europe*, Cambridge: Cambridge University Press, 1990, pp. 13 and 17.

p. 261: Image is a composite of drawings from folios 17r–18v of book Post. 110, Biblioteca Nazionale Centrale, Firenze, by permission of the Ministero per i Beni e le Attività Culturali della Repubblica Italiana.

p. 272: Courtesy of the Topkapı Sarayı Müzesi, H.1654, fol. 306r and H.1653, fol. 406v.

pp. 282–83: Adapted from Marshall G. S. Hodgson, *The Venture of Islam: Conscience and History in a World Civilization*, Chicago and London: University of Chicago Press, 1974, volume 2, p. 534.

pp. 296–97: adapted from D. E. Dumond, *The Eskimos and Aleuts*, London: Thames and Hudson, 1977, pp. 22–23; R. McGhee, *Ancient People of the Arctic*, Vancouver: UBC Press, 1996, pp. 77, 89, 97, 228; Jared Diamond, *Guns, Germs, and Steel: The Fates of Human Societies*, New York: W. W. Norton, 1997, p. 341; G. Jones, *A History of the Vikings*, Oxford: Oxford University Press, 1984, p. 271; M. Chambers et al., *The Western Experience*, New York: Alfred A. Knopf, 1987, p. 457; F. W. Mote and D. Twitchett (eds.), *The Cambridge History of China*, volume 7, Part 1, Cambridge: Cambridge University Press, 1988, p. 234.

p. 317: Courtesy of the C. V. Starr East Asian Library, Columbia University.

p. 322: Courtesy the Library, American Museum of Natural History, Cat. 41.2/4721 and 41.2/4722, from C. M. Torres, *The Iconography of South American Snuff Trays and Related Paraphernalia*, Motala: Göteborgs Etnografiska Museum, 1987, plate 12.

p. 323: (*Left*) Musée des Arts Décoratifs, Paris, photograph by Laurent Sully Jaulmes, all rights reserved; (*right*) Princeton University Art Museum, bequest of Col. James A. Blair, photograph by Bruce M. White.

p. 340: Courtesy of the author.

p. 351: Courtesy of NASA/JPL/Caltech.

INDEX

Page numbers in *italics* refer to illustrations and maps.